D0378982

NOTE-BY-NOTE COOKING

Arts & Traditions of the Table

Arts & Traditions of the Table
PERSPECTIVES ON CULINARY HISTORY

ALBERT SONNENFELD, SERIES EDITOR

LIST CONTINUES ON PAGE 256

Hervé This

TRANSLATED BY M. B. DeBEVOISE

NOTE·BY·
NOTE
COOKING

THE FUTURE
OF FOOD

COLUMBIA UNIVERSITY PRESS   NEW YORK

Columbia University Press

Publishers Since 1893
New York   Chichester, West Sussex

Library of Congress Cataloging-in-Publication Data

This, Hervé.
  [Cuisine note à note. English]
  Note-by-note cooking : the future of food / Hervé This ; translated by M. B. DeBevoise.
    pages cm — (Arts and traditions of the table)
  Translation of: La cuisine note à note.
  Includes bibliographical references and index.
  ISBN 978-0-231-16486-3 (cloth : alk. paper) — ISBN 978-0-231-53823-7 (ebook)
  1. Food additives. 2. Artificial foods. I. Title.

  TX553.A3T4613 2014
  641.3'08—dc23

                                                           2014007035

JACKET DESIGN + BOOK DESIGN + COMPOSITION BY VIN DANG

"

IL N'EST PAS NÉCESSAIRE D'ÊTRE LUGUBRE POUR ÊTRE SÉRIEUX."

HERVÉ THIS

CONTENTS

A NOTE ON
THE TRANSLATION

▼

THE ENGLISH VERSION of *La cuisine note-à-note* differs from the French edition in several ways. At the American publisher's request, and with the author's approval, the contents of the original book have been reorganized so that there are now seven main chapters rather than twelve, and a list of further readings has been included at the end of each chapter. The Columbia edition also contains twenty-nine black-and-white illustrations and a section of twenty color plates not found in the book published by Belin in Paris.

A few minor errors of fact have been silently corrected, and here and there biographical and other details have been added as a courtesy to nonspecialist readers. On behalf of the Press I am pleased to thank Hervé This for his cheerful cooperation in preparing this revised and expanded edition of his work.

—M. B. DeBEVOISE

TABLES, FIGURES, AND COLOR PLATES

▼

TABLES

FIGURES

COLOR PLATES

NOTE-BY-NOTE COOKING

*WHY THE NEED FOR **NOTE-BY-NOTE COOKING** SHOULD BE OBVIOUS*

NOTE-BY-NOTE COOKING? Trust me: I always use words with due regard for what they mean—but I don't deny myself the luxury of metaphor. In the phrase "note-by-note cooking," the noun that is modified is *cooking*. In French, the word *cuisine* denotes a room, the kitchen, but it refers above all to an activity, cooking, well described by the title of a book I wrote a few years ago that was published in English under the title *Cooking: The Quintessential Art*. The original title is more revealing: *La cuisine: C'est de l'amour, de l'art, de la technique*. By this I mean that the cook's primary purpose is to show love (which may or may not be wholly sincere—though the reader will have guessed that I myself favor honesty in matters of the heart) to his or her guests. Beyond that, the cook seeks to make something good, a work of art in the best case, and to make it well, according to some standard of technical excellence.

Let's begin with the simplest part, the technical aspect. Culinary activity physically assumes the form of a series of operations: cutting, heating, grinding, filtering, evaporating, assembling, melting, blending, emulsifying, mixing, expanding—exactly the same things, as it happens, that are done in the marvelous field of chemistry. There is nothing in the least surprising about this: chemistry, like cooking, has a technical component. But this component is almost an incidental detail. Chemistry, though it involves technique, is not merely a matter of "doing." It is also a science, a search for the underlying mechanisms of phenomena whose aim is to acquire knowledge. What does it mean to say, then, that cooking has a technical component? It means that cooking is not solely a matter of technique, as most people suppose. There is much, much more to it than that.

If cooking were only a matter of technique, it would be a very sad thing indeed, for a machine is all that would be needed to cut, heat, grind, filter, evaporate, assemble, and so on. Worse still, machines often outperform human beings from the technical point of view. But cooking is able to transcend mere doing by virtue of the fact that it also has an artistic aspect. It aims at making something good, which is to say dishes that are fine and beautiful things to eat. Technique is a means, not an end. And so we are agreed that cooking is both a technical and an artistic activity. But that is by no means all it is. Even the most beautiful, the most splendid meals are not always good because we partake of something more than food when we eat. We partake also of the love shown to us not only by the cook, but also by those with whom we share a meal. It is often (and rightly) said that if you sup with the devil, you need a long spoon. This is why expensive business meals are apt to be so unsatisfying, whereas a simple snack with good friends can be one of the loveliest experiences there is. We do not eat the way animals do. We are social creatures, endowed above all with the power of speech, and it is the act of gathering together (some food lovers even speak of "communion"), in which eating is combined with conversation, that accounts for the happiness of so many moments we spend at the table. This is what La Varenne meant when he said that jolly meals are the best ones of all.

My use of the term *love* to describe the third aspect of cooking (in reality, the first in order of importance) is no doubt inappropriate, at least some of the time, for we do not all look at it in the same way. For some people, love really is love, but for others it is power or money or something else. The command "Eat your vegetables!"—unfondly remembered by many of us from childhood—is an example of the power exerted over us by our parents, not always benevolently, when we are young. When we are older, doctors and nutritionists order us to eat in ways that they consider to be healthy. The word *love* may seem a bit naive, I grant you, but I use it deliberately, even if it is clearer, more reasonable, to speak of cooking as having a social aspect. And obviously I don't mean to suggest that we should want to eat in ways that are unhealthy. We are right to prefer that cooking be done by trained professionals rather than run the risk of being poisoned by self-taught amateurs. But in that case we put ourselves on the other side of the cooking line, where the waiters are loading up their trays with the dishes that they are about to bring out to us in the dining room.

So much, then, for the "cooking" part of "note-by-note cooking." The "note-by-note" part is a bit trickier. We will need to examine it with some care.

A NONOBVIOUS PIECE OF OBVIOUSNESS

It is curious that the value of note-by-note cooking is not thought to be plain on its face, but instead becomes apparent only after one has thought about it for a while (which says a great deal about what is or is not obvious to us, but that's another story). And yet we shall see that note-by-note cooking altogether naturally recommends itself to cooks and gourmets alike.

We cannot avoid talking in the first place about the love of food, for our subject is cooking. If we love wild mushrooms, for example, we are naturally saddened by the thought that the season for them will eventually come to an end and they will become vanishingly scarce. It makes us melancholy to see the forest floor fill up with water and then freeze, to see the ground covered by dead and decomposing leaves, a sad brown carpet on which our marvelous specimens are nowhere to be found. The only way to lift our spirits is to buy imported fresh mushrooms or to use products that have been preserved in one or more ways, ancient or modern, from dried mushrooms to canned. Many kinds of food are preserved, of course (candied fruits, pickled vegetables, conserved meats, and so on), to say nothing of the great variety of frozen foods available to us today. Nevertheless what is scarce is expensive, and we do not always feel wealthy enough to buy everything we would like.

Then one evening a friend comes over for dinner—a true friend, as it happens, since he solves our problem. He tells us about 1-octen-3-ol, a compound that is found in many mushrooms. We listen to him with an open mind because gastronomy is part of what makes us human: by helping us to go beyond our animal instincts, it has taught us not to rule anything out without subjecting it first to thorough examination. Gastronomy teaches curiosity, an essential element of both the arts and the sciences. The enlightened food lover in us therefore seeks to know more about this compound with the strange name. We get hold of a sample, smell it, and discover that 1-octen-3-ol (or octenol, for short) has a truly marvelous fragrance, very similar to the ones that we love and miss so much. To be sure, it is not quite the same as the odor of a chanterelle or a cèpe or a black trumpet. No, the scent is familiar and novel at the same time: in addition to the smell of mushroom, there are notes of undergrowth, the undergrowth of a deep and wet forest. What should we do? Should we be content with only a memory—the memory of our favorite mushrooms? Should we go on living, in other words, in the realm of the mind? Or should we embrace the sensual appreciation of good food and get ourselves into the kitchen?

The true gourmet, once tempted, cannot resist. Quickly he grows accustomed to cooking with this new product, just as he would get used to cooking with a new spice (spices, by the way, contain a great many flavorful compounds, most of which we are unaware of). Having come to feel comfortable using this first compound, the enlightened food lover now goes on to acquaint himself with a second one, limonene, which has a wonderfully fresh scent that calls to mind the fragrance of citrus fruits such as lemon and orange. And then a third one, sotolon, which, in addition to notes of walnut, curry, and fenugreek, has an odor reminiscent of the famous "yellow wine" of the Jura in France. And a fourth, arginine, which, although it has no odor, yet has a very distinctive flavor.

The gourmet's scruples melt away in the face of so many intriguing sensations. Familiar products such as rose water, orange blossom water, nutmeg, parsley, and ginger are soon joined in his kitchen pantry by a variety of chemical compounds in liquid or powdered form. Before long the pleasure he derives from cooking, which inevitably comes before the pleasure he takes in dining with others, leads him to mix two of these novel culinary preparations, then three of them, then four. Fruits, vegetables, meats, and fish eventually disappear, giving way to pure compounds, which are skillfully combined to create dishes of a new kind. And there it is—note-by-note cooking!

THE HISTORY OF NOTE-BY-NOTE COOKING

The just-so story I have just imagined is not so very different from the actual history of the creation of note-by-note cooking. It all began in 1994, when I was rushing to finish an article for *Scientific American* entitled "Chemistry and Physics in the Kitchen" with my dear friend Nicholas Kurti. He was fifty years older than I was, but we were like two children for whom the difference in age counted for nothing next to our desire to do wonderful things and, even better, to do them together. Nicholas was a remarkable physicist. He had had an eventful life, leaving Budapest, where as a young man he had contemplated a career as a concert pianist, for Paris, then moving to Berlin, and finally to Oxford after being driven out of Germany by the Nazis. He is remembered in particular for the discovery of nuclear adiabatic demagnetization, a technique he used to achieve the lowest temperatures ever recorded, only fractionally above absolute zero (–273.15°C)—the temperature below which it will never be possible to go, for it is at that point that atoms cease to move or change position.

Six years earlier Nicholas and I had created the scientific discipline that came to be known as "molecular gastronomy," taking advantage of every possible opportunity to work together. When I was asked to set up an institution of higher learning in France dedicated to food research, I enlisted Nicholas's assistance as cofounder. When Nicholas was asked to write scientific notes for a cookbook, he sought my aid. Thus it was that we came to collaborate on the article for *Scientific American*.

For such articles, there was one invariable rule: I wrote the first draft, which Nicholas then edited. It was an efficient division of labor. For the *Scientific American* piece we had agreed on an outline in advance, so the writing went quickly. It remained only for me to come up with a conclusion. But what could I say in conclusion when the adventure of molecular gastronomy had scarcely begun? At the time my own work had less to do with octenol than with another volatile compound, paraethylphenol, which in very weak

SCIENTIFIC AMERICAN

APRIL 1994
$3.95

The dilemmas of prostate cancer.
The real culprit in U.S. economic ills.
Watching the Mind at work.

Ancient Peruvian mask and headdress offer
clues about a mysterious pre-Incan civilization.

0.1 THE COVER OF THE APRIL 1994 ISSUE OF SCIENTIFIC AMERICAN *IN WHICH THE PRINCIPLES OF NOTE-BY-NOTE COOKING WERE FIRST SKETCHED. REPRODUCED WITH PERMISSION. COPYRIGHT © 1994 SCIENTIFIC AMERICAN, INC. ALL RIGHTS RESERVED.*

concentrations contributes to the smell of leather given off by vintage Burgundy wines and, in only slightly higher concentrations, transforms an ordinary whiskey into a peated beverage. Since I was busy investigating the properties of paraethylphenol in my lab, it occurred to me to tell our readers what I had found out so far. One thing led to another, altogether naturally and logically it seemed to me, with the result that the idea of note-by-note cooking was born. Nicholas was willing to indulge me—partly because the story I had to tell was not entirely uninteresting, but mainly because we had an urgent deadline to meet! Even so, and even though I was a chemist, I worried that I was pushing things a bit too far. Was I really suggesting that we should add compounds to these marvelous works of art known as wines? Indeed, no sooner had the article appeared than oenophiles rose up in protest. If this sort of thing were permitted, they said, the world of wine would be fatally corrupted. The myths of Pandora and Prometheus combined to raise the specter of a future in which science is used to encourage first the doctoring and adulteration of wines and then the marketing of them by every known means of fraud and dishonesty.

Are wine lovers' fears justified? We will never be able to reply to this question, or to any other question worth asking, if we do not take care to speak as precisely as possible. Adulteration—making wine impure by adding alien ingredients—is against the law. The modification of wine is not in and of itself a serious matter. What is a serious matter, from the legal point of view, is giving the name *wine* to something that plainly is nothing of the sort. By definition, wine is a product that can be obtained only through the fermentation of grape juice. From this it follows that adulteration is improper. How is it, then, that we have come to accept the idea, common today, that adulteration may be a source of improvement? I fear, alas, that "sophisticated" women are at least in part to blame because their love of cosmetics

0.2 SOME PARAETHYLPHENOL CRYSTALS.

makes them victims of adulteration (the literal meaning of the French word *sophistiqué*). In our culture, however, women who apply makeup are considered beautiful. Surely this is why the idea of adulteration has gradually acquired a meliorative connotation, without due regard for what words mean.

The fact remains that doctoring and adulteration, particularly of wines, are unlawful activities. This, as I say, is because the law does not authorize use of the word *wine* to designate anything other than fermented grape juice. The addition of sulfur is permitted so that the wine will keep longer, but nothing else. And yet, in my view, even the use of preservatives should be opposed. There is nothing wrong, of course, with making wine-based products (as indeed the ancients did by adding honey and various aromatics), but in that case we are no longer dealing with wine; we are dealing instead with a product that is made from wine and that must therefore be labeled as such. The same is true of bread, which by law can be made only from water, flour, yeast (and, again, a very restricted number of additives): bread must be bread—not cake or any bread-based preparation. New preparations need new names. There is no reason to debase the names we have been using correctly all these years.

Note-by-note cooking, I am pleased to say, has nothing to do with any of this. It is not a method for altering or adulterating or doctoring foods. It is a method for making foods out of compounds.

NOTE-BY-NOTE COOKING IS NOT "CHEMISTRY"

From the very beginning, then, there was adamant opposition to note-by-note cooking on the ground that putting "chemistry" into cooking would be a terrible thing. Why the scare quotes? Because chemistry will never be a part of cooking! It is an impossibility. Chemistry is an activity that seeks to acquire knowledge: it seeks to understand how atoms are arranged in matter and how they can be rearranged in the course of reactions. Chemical reactions? No, reactions. Atomic reactions, perhaps, because atomic reactions are what produce molecular compounds—but not chemical reactions. The expression *chemical reaction*, like *chemical product*, should be used only very seldom, as we shall soon see.

First, however, let me emphasize that however shocking the claim that chemistry will never have a place in cooking may seem, it is nevertheless perfectly true. It is not a piece of linguistic legerdemain palmed off by a chemist looking to defend the parochial interests of his profession. To see why, let's begin by taking a closer look at the class of substances known as compounds—substances so poorly understood that they are commonly confused with molecules, products, chemical products, synthetic compounds, and who knows what else. The easiest way to do this is to step into the kitchen. Put a piece of beef sirloin in a hot oven, say 200°C (almost 400°F). After a few minutes, the surface of the meat will have turned brown because compounds in meat that are heated to more than 100°C (212°F) react and are transformed. Are these chemical compounds? No, they are compounds. Water is a compound because it is made of very small objects, constantly in motion, that are called molecules (in this case, water molecules). If water is a compound, it is not because it is composed of water molecules, but because water molecules, all of them identical, are composed of atoms of different sorts. In a water molecule, for example, there are two atoms of one sort, hydrogen, and one atom of another sort, oxygen. What are atoms? Let us content ourselves for the moment with saying that they are objects smaller than molecules since molecules are composed of atoms.

Molecules make up the main part of living matter and therefore of our bodies as human beings and of our environment. In wines, for example, the principal molecular compounds are water, ethanol, and tartaric acid. In the starches obtained from potatoes and rice, the compounds are amylose and amylopectin (not to be confused with the pectin found in jams and jellies). In oils, the compounds are triglycerides. An egg white is composed chiefly (90 percent) of a single compound, water, along with smaller amounts of various other compounds known as proteins. There is nothing in the least "chemical" about these compounds. The confusion arises in part from a logical error concerning the relation between a whole and its parts. Referring to the train of officials following a head of state, for example, as a "presidential procession" is an error of this type because the procession could be presidential only if it itself were the president. Similarly, a compound is a compound. It becomes a *chemical* compound only when it is studied by chemists.

The mistake in talking about "chemical compounds" involves a deeper confusion, however, between so-called natural and synthetic compounds. The difference between them will become clear at once if we look more closely at what water is made of. A drop of water that falls from the sky is composed of molecules, each of which has two hydrogen atoms and one oxygen atom. These molecules (all of them of natural origin because the sky is a part of nature, not a product of human activity) are in every respect identical to those that a chemist would obtain by making two gases—dihydrogen (one used to say simply "hydrogen," but there was a risk of ambiguity, which modern chemistry has since dispelled) and dioxygen (formerly called "oxygen")—react : a flame or a spark that comes into contact with a mixture of these two gases produces an explosion, with the result that the atoms of the two gases are reorganized into water molecules (if this makes you nervous, rest assured that chemists today know how to synthesize water without the gas exploding). Whether it is water that rains down from the sky or water that has been synthesized by means of a reaction between dihydrogen and dioxygen, it is made up of molecules, and each of these molecules consists of a unique combination of three atoms, two of them hydrogen and one oxygen. The physical, chemical, sensory, and nutritional properties are entirely the same for the two sorts of water. And yet one of the two has been synthesized, the other not.

The same thing is true of acetic acid, the compound mainly responsible for the acidity of vinegar, and of vanillin, the principal compound of the smell of vanilla. Here again one finds molecules that are wholly identical to artificial molecules made in a laboratory. In these and many other cases, the operations carried out by the chemist are indistinguishable from what a cook does when he cooks. Why should we be bothered then?

Chemists also know how to make compounds that are not found in nature. By rearranging natural compounds, they gradually learned to construct compounds that had not been previously identified. This does not mean that such compounds do not in fact exist, only that they are not known to exist in nature, at least not for the moment, and that there is no real point trying to find out whether they do already exist, unless perhaps to settle trademark or copyright disputes, for example. This, at any rate, is what most chemists would tell you. Many nonchemists fear that fabricating novel compounds would only lead to unforeseen troubles. But once again it needs to be pointed out that cooks do exactly the same thing when they impart flavor to food by cooking it: the heating causes compounds to appear that initially were not present and that in many cases do not (so far as we know) exist in nature. No one seems to think this poses a problem.

Whether the nonchemists are right or not, we all can agree that a synthetic compound (one should really speak instead of a "synthesized" compound or of a compound that has been made "by synthesis") is a substance produced by human beings through the reorganization of matter brought about by a reactive process. Again, rather than speak of a chemical reaction, we should refer instead to a rearrangement of atoms—a slightly more cumbersome phrase, I admit, but how much more accurate! At all events it should be clear that chemistry will never be found in cooking. At worst (or at best, depending on your point of view) there will be—and already are—synthesized compounds in addition to so-called natural compounds.

NATURAL FOODS? IMPOSSIBLE!

In distinguishing between compounds of natural origin and compounds that have been synthesized in a laboratory, I passed too quickly over the idea of nature and what it is supposed to signify. In an era of self-righteous environmentalism, these things are inevitably a crucial part of the debate that note-by-note cooking has helped to inaugurate in the world of food.

Whole treatises have been devoted to the subject. All of them begin from a true premise—namely, that a thing is natural rather than artificial by virtue of the fact that it has not been transformed in any way by human intervention. An exploding volcano is a natural phenomenon, for example, whereas a musical melody is an artificial creation. From this it follows that no prepared food is natural, for prepared foods, by definition, are the result of human preparation and therefore artificial from the very start.

Moreover, most traditional (as opposed to note-by-note) dishes are made from animal and vegetable products that themselves are artificial since they have been modified to one degree or another by human activity. Just as it

0.3   *THIS VOLCANIC ERUPTION IS HORRIBLY NATURAL.*

takes a gardener to grow carrots, a breeder and a butcher are needed to bring a leg of lamb to our kitchen.

In other words, because even the ingredients used in traditional cooking are not natural, foods are still less natural than one might at first be inclined to suppose. (Indeed, a scale of artificiality could be devised to measure the degrees of separation between a product and pure naturalness.) Complicating matters still further, modern culinary ingredients are typically parts of animals or vegetables that themselves are the result of a long process of selection. Consider, for example, how great a distance separates the natural wild carrot, a hard and fibrous stick, from the artificial cultivated carrot, bright orange and swollen with sugars. Our foods are not natural—and they never were, no matter how stridently natural-foods advocates and others may insist on it.

This distinction has now become an issue of public policy, at least in the European Union, where regulatory authorities are being pressed by various interest groups (government agencies, food companies, consumer groups, ordinary citizens) to provide a rigorous definition of what constitutes a "natural" food. Evidently this is an absurdity, for the reasons I have just given.

0.4   *THIS DISH BY PIERRE GAGNAIRE IS MARVELOUSLY ARTIFICIAL.*

The same absurdity will present itself later, in chapter 4, in connection with odorant compounds. A most unwelcome regulation has recently authorized the use of the term *aroma* to designate various combinations of odorant compounds, whereas plainly it should be restricted to the scent or fragrance of aromatic plants—aromatics, as they have long and rightly been known. Companies that prepare and market such mixtures (some of which are terrific, by the way) should not be granted a monopoly on the term *aroma*. We mustn't yield to demagogy or commercial pressures, even if they have the force of law: if a law commands us to believe that 2 + 2 = 5, it is our duty to oppose it.

Conversely, it must be recognized that the term *artificial* is derived from *art*. The artificial is something quite remarkable, far superior to the natural, because it has been designed by human beings and realized through their own labor. Our clothing, our homes, our computers, our medicines and protective cosmetics—all of these and many more things are artificial. Without them, most of us would have been dead a long time ago. To be sure, not all things made by human hands are good—or, shall we say, equally good. A bomb, for example, is a lethal device; rape, murder, and other violent crimes are abhorrent. There are, of course, grades of quality, just as there is a difference in quality between the drawing made by a child who has not yet learned how to hold a pencil and a drawing by Dürer, who spent a lifetime mastering his art. But even if we may agree that not all artificial things are equally good, isn't it up to us to try to improve them, to make them better?

People who talk of a "golden age" of food—an age when ingredients were natural, when cooking was better than it is now, and so on—are perpetuating a myth, a childish illusion that we are loath to abandon. Someone recently wrote me that because she had been raised in the countryside, her parents and grandparents made sure she had a wholesome diet. I do not doubt that she was raised in the countryside, if she tells me she was, but I am rather skeptical about the expertise of parents and grandparents in this matter. What did French peasants really know about diet in the early twentieth century? Can we really say that their own diet was wholesome, considering how many people in rural villages in France died at a relatively young age in those days? Without modern advances in epidemiology and nutritional science, we would still be consuming enormous quantities of smoked products that cause cancers of the digestive system; we would still be adding cinnabar (mercuric sulfide, a horribly toxic substance) to *pissala*, the paste traditionally prepared in Nice from anchovies and salt; we would still be sweetening wines, as the Romans did, with intoxicating lead oxides. When it comes to diet, the wisdom of the ancients may be doubted.

Ought we not, in fact, take the view that the ancients are not to be listened to at all, no more in matters of cooking, science, or art than in matters of nutrition, and that it was the cumulative labor of generations that led to an increase in knowledge and humane learning? The primitive understanding of pictorial perspective in Egypt, under the pharaohs, when figures were shown in profile, underwent no fundamental modification until Leon Battista Alberti published his treatise *On Painting* in the mid–fifteenth century. Plato, one of the greatest of all philosophers, attempted to justify slavery. The classical doctrine of the inferiority of certain human races continued to be taught as late as the twentieth century. Surely there can be no good reason not to free ourselves at long last from this idea that everything that is ancient is good?

The question whether the artificial is better than the natural (or vice versa) makes no sense because it is too general. We need to recognize that what is naturally good—Caesar's mushroom (*Amanita caesarea*), for example—is good, but what is naturally bad—the death-cap mushroom (*Amanita phalloides*), for example—is bad; likewise, that what is artificially good—Bach's *Sonata for Solo Flute*, for example—is good, but what is artificially bad—a very cancerous compound, for example—is bad. Moreover, even though nothing is good or bad in an absolute sense, there must nevertheless be an apt correspondence between a thing and its function. A knife, useful when used to prepare food, may be an instrument of evil when used to commit a crime.

Why do some people believe that the artificial is unequivocally bad? No doubt because we are animals who have evolved over time, on the one hand, and because not all the things we make are equally good, on the other. There is no disputing the first claim: we are indeed animals, and as animals we have been "coded" by billions of years of biological evolution. If a newborn child (no less than a newborn monkey) smiles when sugar is rubbed on its lips, for example, it is because its animal instinct tells it that sugar is a source of energy. This is not surprising since primates coevolved with plants: the sweet fruits produced by plants attracted primates, and primates dispersed those plants' seeds in turn. As for the second claim, that not everything that is artificial is equally good, nothing could be easier than to multiply examples beyond the ones I have already given.

Take the case of vanilla, vanillin, and the various extracts and preparations that are sold in grocery stores. To begin with, vanillin is a compound and must not be confused with vanilla, a fermented pod. Keep in mind, too, that vanilla is not in the least natural. It is the result of an elaborate manufacturing process in which the pod (or capsule) of an orchid of the genus *Vanilla* is subjected to a series of procedures: scalding, sweating, drying, conditioning (during which it is kept in closed boxes and cured for five to six months), and grading. The capsule turns brown only during the sweating stage, when an enzymatic reaction occurs, and it releases its aroma (the proper term in this case since vanilla is an aromatic plant) only during the drying and conditioning phases, when the moisture content is reduced.

Much of the flavor of vanilla in its processed form is due to the compound vanillin, but the richness of the flavor, its mellowness, results from the presence of many other compounds as well. Vanillin is only one of the compounds found in vanilla. It accounts for a part of the flavor of the fermented pod, but not all of it. Moreover, it is false to speak of *the* flavor of vanilla because no two neighboring pods have exactly the same flavor. Just because trained tasters cannot tell the difference when they are asked to compare vanilla and vanillin, should we allow the two things to be conflated? No, a thousand times no! That would be a lie—and all the more so since vanillin can be extracted from lignin, a compound found in wood, which is much less expensive than the plant from which vanilla is harvested. Is vanillin somehow better than vanilla, or is it less good? There is no clear answer, for "better" in this case is a matter of individual preference. However it is correct to say that vanillin is vanillin (whether it is natural or synthetic) and that vanilla is vanilla (a quite artificial product, considering the number of steps required to make it).

It must be admitted that not all preparations that give off a scent of vanilla are equally good. Some, prepared by gifted artists who have spared no expense, are very fine; others, in which vanillin has been dissolved in an ordinary caramel, for example, are unmemorable. Whether the vanillin is natural or synthetic, however, is of no importance whatsoever. We all can agree, I think, that what matters is the final result.

COMPOUNDS AND MOLECULES

Compounds are evidently a stumbling block for many people. They will feel more at ease with the idea of note-by-note cooking once they understand what compounds are, because compounds are at the heart of note-by-note cooking. The way forward will be to take a closer look at the notion of molecular structure. Let's start by considering some unfamiliar solid object. No, wait, let's start with water—because no one argues about it (except, of course, gourmets, who are also lovers of wine). Better yet, let's start with a glass of water. There is something wondrous about even this, for water is liquid, whereas the glass that contains it is solid. Water is transparent, whereas many physical objects are opaque. Water is odorless, whereas many flowers have a smell. Water is—

Hold it right there! We don't care about water in and of itself, only about using it to illuminate a fundamental fact of the physical world. Let's confine our attention for the moment to the water that's in this particular glass and divide it into two parts: each part is likewise water. Divide each part again. We still have some water, less than before in each fraction, of course, but it's still the same: still colorless, still transparent, still liquid (at least at room temperature in the temperate zones of the world). Now go on dividing again and again and again. So long as we divide fewer than eighty-three times, we will still have some water: less and less each time, until it can no longer be seen without specialized equipment, but water just the same.

Then, suddenly, at the eighty-fourth division, the water is no longer water, but another form of matter. It is at this point that we go beyond the smallest part that can be called water: the molecule. There is no need for the moment to go further, to the level of atoms, the constitutive elements of molecules. For the moment, it is enough to keep in mind that a "pure" substance is made of identical molecules. Water is made of a great many water molecules. Indeed, there are hundreds of thousands of millions of billions of them in a single glass.

The same is true for ordinary sugar, which chemists call "sucrose." Divide a lump of sugar in two, and each of the halves is still sugar; divide it again and again, and you still have sugar—until the moment when, to the taste, the sensation of sweetness finally disappears, even for an ideal palate. The last stage of division in which the tiniest bit of sweetness can still be recognized, assuming a faculty of perception that is extraordinarily sharper than our own, is that of the molecule. The same goes for salt, for acetic acid, for ethanol, for limonene, for citric acid, for hydroxyproline, and so on.

Salt and sugar are very familiar to us, but what about the other items in this list? Unlike salt and sugar, they are not used in pure form in traditional cooking. Instead they are mixed with other compounds. I have already mentioned acetic acid. It is the principal compound of vinegar, the thing that gives vinegar the main part of its acidity. Once again, the best thing will be to start from what we know. The known in this case is white vinegar (vinegar, that is, whose color has been removed by distillation or filtration). In this liquid, just

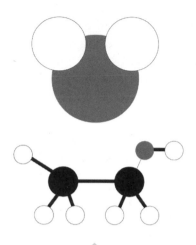

0.5 *A WATER MOLECULE CAN BE REPRESENTED AS A MICKEY MOUSE HEAD, AND AN ETHANOL MOLECULE AS A SMALL DOG.*

as in water, there are molecules. Nine out of ten of them are water molecules; the others are mainly (though not exclusively) acetic acid molecules, without which white vinegar would have little to recommend it. Molecules of other compounds are present in such small quantities, in fact, that for present purposes we may as well think of acetic acid as being composed simply of water molecules and acetic acid molecules.

These various molecules display both similarities and differences. Let's start with the similarities. All molecules (more precisely, all molecules that make up compounds) are formed of atoms. We may think of atoms as little billiard balls of different sizes and colors. Not that atoms really are billiard balls or that they really have colors. I use the image of billiard balls of different sizes to make it easier to picture them and colors to make it easier to tell the different atoms apart. These atoms have names: *hydrogen, oxygen, carbon*, and so on. There's nothing very complicated about any of this. There exist in nature only about a hundred distinct atoms, known as "elements."

A water molecule contains only three atoms, two of hydrogen and one of oxygen. Again, to make things easier to visualize, we may think of molecules as resembling Mickey Mouse heads. No offense is intended to chemistry by this comparison, of course; it is meant only as a way of getting our minds around an unfamiliar idea. (Besides, chemistry couldn't possibly take offense—it's a science, not a person!) In an ethanol molecule, unlike a water molecule, a carbon atom is attached to three hydrogen atoms and to one other carbon atom, which in turn is attached to two hydrogen atoms and one oxygen atom, which is itself attached to a hydrogen atom—rather like a small dog. And so on. As I say, there's nothing very complicated about any of this. It's simply a matter of attaching billiard balls—atoms of various sorts—to one another in order to make different molecules.

Here it occurs to me once more that words often get in the way. Because thought is inseparable from language, we have a hard time thinking without the right words. In chemistry, there is no way to avoid having to learn a few basic terms: *molecule, compound, substance, matter,* and so on. Not the least of the reasons why nonchemists are liable to be disconcerted by the word *compound* is that it designates a category, which is to say an abstract concept. And yet just as we are in the habit of designating every tree that belongs to the category of oaks by the word *oak*, we can do the same thing with different kinds of molecules—but with this difference: we do not call an acetic acid molecule "acetic acid"; we call it an "acetic acid molecule," which has the advantage of allowing us to distinguish between a particular object and the class of such objects.

It is true, of course, that few of us have ever seen acetic acid or citric acid or hydroxyproline or indeed any other compound in its pure form. But this is nothing to be concerned about. In their pure form, these compounds are generally powders, crystals, or liquids. They are typically white, but not always. Beta-carotene, for example, one of the compounds that contributes to the color of carrots, is bright orange; pure chlorophyll *a*, which is present in most green vegetables, is—you guessed it—green. And so on.

FROM MOLECULAR COOKING TO NOTE-BY-NOTE COOKING

We will have a better idea of what note-by-note cooking involves if we place it in its historical context. For this purpose, we need to proceed in much the same way as painters do, adding to our canvas little touches of color that gradually combine to form a coherent image. To begin with, a few definitions that I have come to realize are indispensable, not least in trying to dispel the persistent confusion between cooking and gastronomy:

- ⊞ Cooking is an activity that consists in preparing foods in the form of dishes.
- ⊞ Gastronomy is the intelligent knowledge of whatever concerns man's nourishment, as Jean-Anthelme Brillat-Savarin put it in *The Physiology of Taste* (1825)—a definition that has not been improved upon since. Brillat-Savarin popularized the term *gastronomy*, which had been introduced by the poet Joseph Berchoux a few years earlier, at the turn of the nineteenth century.
- ⊞ Molecular gastronomy, as more and more people are beginning to realize, is a branch of physical chemistry that explores the results of culinary transformations. In the extreme case, it has no interest in the activity of cooking itself and looks upon the chemical art embodied by cooking solely as an occasion to discover novel phenomena and mechanisms.
- ⊞ Molecular cooking is a way of cooking that makes use of new utensils, ingredients, and methods. Here the word *new* is to be understood as referring to anything not found in the cooking of, say, Paul Bocuse. Molecular cooking was created in the early 1980s, when I had the idea of adapting for use in the kitchen all (or almost all) of the devices and materials found in a chemistry laboratory: rotary evaporators, ultrasound probes, liquid hydrogen, separating funnels, filter pumps, and so on. What began as an eccentric culinary tendency in France has by now become a global phenomenon: not a day goes by without news of yet another chef who has been converted to this style of innovation. It is evidently the result of applying science to technology. Just the same, one must hope that molecular cooking will soon die out—though not, of course, molecular gastronomy! The obsolescence of molecular cooking would be proof that a decisive technological transition has been accomplished—that the activity of cooking has at last been modernized.
- ⊞ Note-by-note cooking was first imagined in the mid-1990s. The following passage occurs near the end of the *Scientific American* article I mentioned earlier:

The manufacturers of wines and spirits are typically forbidden by law to improve the taste of their products by adding sugar or other chemicals.

Nevertheless, if the consumer wants to use the results of chemical research to enhance the qualities of inferior wines or spirits, should he or she not be encouraged to do so? A few drops of vanilla extract may wonderfully enrich the flavor of a bottle of cheap whiskey.

This kind of experiment can be extended to a large number of beverages and dishes. Perhaps in the cookbooks of the future, recipes will include such directions as "add to your bouillon two drops of a 0.001 percent solution of benzylmercaptan in pure alcohol."

The idea wasn't crazy in the least. Cooks have always modified the flavor of their dishes by adding herbs and spices, which are nature's way of packaging mixtures of aromatic molecules. At all events, as I say, Nicholas Kurti approved this passage, although I myself considered it to be a piece of extravagant audacity. What? Chemicals in foods? Today I would state the matter differently—mainly, of course, because the term *chemicals* (like *chemical compounds*) is mistaken: a compound is a compound, whether it is extracted from a natural or a synthesized product; it becomes a chemical only if it is used or studied by a chemist. The water we drink, for example, is a chemical compound only insofar as it is the object of chemical analysis.

I also suggested that cooks might find it entertaining to experiment with compounds. But entertainment is not really the point, or at least not the only one. A few drops of vanillin (whether synthetic or natural hardly matters because the molecules are the same and have the same effects on sensory receptors) do, in fact, give roundness to unaged alcohols. The reasons it does so are, to the best of my knowledge, unknown. The phenomenon has something to do with the fact that during the barrel aging of brandies the lignin in the wood reacts with the ethanol and over time produces a number of compounds, including syringaldehyde as well as sinapic and vanillic aldehydes. For a lover of brandy who is not rich enough to buy expensive brands that have spent many long years in cask, the appeal of adding vanillin should be obvious.

Finally, I now believe that even if adding benzyl mercaptan to bouillon is a good choice from the culinary point of view (in solutions of low concentration, benzyl mercaptan has notes of onion, garlic, horseradish, mint, and coffee), there is a risk that my fellow chemists, who think of mercaptans as volatile sulfur compounds, will balk. In 1994, however, it seemed to me that the idea of adding compounds to foods did not go nearly far enough. Why not *make* foods out of compounds? With this question, note-by-note cooking was born.

Experience nonetheless has proved that people have a hard time imagining what note-by-note dishes might look like. In the pages that follow, I try to show how they can be constructed. Because cooking is an art, however, let me first suggest a comparison with two other arts, music and painting, in the hope that even the most reluctant readers will find the idea of note-by-note cooking less off-putting.

In the past, music was made with instruments—string instruments, wind instruments, and so on. The trumpet, the violin, the piano, and other instruments were used alone or in combination with one another when greater variety was desired. Each instrument had its own timbre. Each century added its own style of composition, its own ways of arranging notes for different instruments, singly or in unison.

With the advent of electronics about a hundred years ago, engineers who saw that the new technology could be used to make music joined forces with musicians who saw in it a new and less cumbersome way to compose and perform music. They were visionaries, not opportunists. Their response expressed a perfectly human desire. As children playing in the park, we feel the urge to go outside it and explore the world beyond. This urge doesn't leave us once we have grown up.

In the 1940s, musicians such as Pierre Schaeffer, Karlheinz Stockhausen, and Edgar Varèse began to mix sounds produced by electronic devices with sounds produced by classical instruments. Electronic music has made enormous strides since then—to the point that today it is everywhere. Using synthesizers to assemble elementary sound waves, it is now possible to produce any sound whatever, whether one seeks to recreate the sounds of classical instruments (but why bother, since these instruments already exist?) or to invent novel sounds.

When we perceive a sound, we perceive an attack, a timbre, a pitch, an intensity. In cooking, when we perceive a flavor, we perceive a consistency, a taste, an odor, a level of heat. The comparison is obvious. And the culinary equivalent of a sound wave is a compound. In cooking, as in music, the arrangement of well-chosen sensory units allows us unlimited freedom: we can produce any sound, any music or any flavor, any food. And, in either case, everything has to be constructed. One might imagine that cooking from molecular scratch, as it were, on the basis of compounds, is somehow more difficult than traditional cooking; but the same reservation in the case of music (it used to be said that instruments without a fixed, predetermined timbre would be very difficult to play) was proved to be unfounded by the imagination of modern composers, acousticians, sound engineers, and musicians themselves.

The construction of a dish from compounds, note by note, is bound to be a long and involved business. But there is no reason why we should not look for shortcuts. The very same thing, after all, is true of perfumes—which themselves are nothing more (or less) than novel combinations of compounds, and not just ones that have been extracted from plants. The same is true as well for the various compositions and extracts (essential oils, resinoids, supercritical carbon dioxide extracts, and so on) that the food industry uses to make fragrances and other synthetic products (the meaning of the word *aroma*, as I say, has been abusively enlarged to include this class of artificial compounds).

Next, a visual comparison. We know that when yellow and blue are mixed together the result is green and that by mixing blue with red we get purple. By contrast, mixing purple and brown together will never yield yellow, blue, and red. These three primary, or basic, colors can be combined in various ways to obtain any color, but no mixing of such combinations will give us back their primary colors. The analogy with note-by-note cooking? Meats, fish, fruits, and vegetables are mixtures of a very great many compounds, and mixing them together creates complex flavors. But it will never be possible to obtain a "pure" flavor by combining mixtures. Combine beef and carrots any way you like, you will never succeed in producing the flavor of lemon.

QUESTIONS

Even people who understand that chemistry will never be a part of cooking are apt to find the idea of note-by-note cooking unsettling. Does it have any nutritional value, they ask? Can we be sure it isn't harmful? What implications does it have for agricultural production? For traditional cooking? For our traditional notions of conviviality?

Their greatest fear is that traditional cooking—cassoulet, pot-au-feu, choucroute, and so on—will disappear. The virtues of these dishes are often praised with much insincerity, considering how little the dishes themselves are really liked! Chocolate, however, is very much liked. But why should we insist that it contain at least a few milligrams of potassium (have you ever seen a milligram, by the way?) when it consists overwhelmingly of fats and sugar (a particular kind of sugar, sucrose)—compounds that no one recommends for their nutritional value? Why should we praise meats grilled over an open fire when they are packed with two thousand times the amount of benzopyrene (a notoriously carcinogenic organic compound) allowed by European law in industrially processed smoked-meat products?

The neophobia that has protected primates through the ages leads us to place our trust in the foods we eat when we are young and to fear new foods. From an evolutionary perspective, neophobia is a salutary reflex because it prevents us from indiscriminately consuming foods that may be harmful, even toxic. Even so, it is very much an animal behavior that we should want to go beyond, or at least to critically examine, so that we cease to be in thrall to it—and all the more since human beings have progressively become omnivores, which means that we no longer run the risk of starving to death when there is a shortage of the food that our nonhuman primate cousins consume almost exclusively: fruits.

How do we go about transcending our animal nature? By symbolizing, ritualizing the act of eating; by rationalizing, finding reasons for eating one way rather than another. Over time we have come to think differently about how we eat. But it is a gradual process. We no longer reject new foods outright, as our primate ancestors instinctively did; still, we disparage new foods and champion familiar foods, even when their virtues are far from being proven. The least persuasive of the justifications given for this is that because familiar foods are ancient, their dangers, if any, would surely have become apparent by now.

Arguments made in bad faith are obviously bad, for exactly this reason. Smoked products, for example, have been legitimized by long-standing custom. Although it is true that smoking was useful over the course of human history as a way of preserving perishable food stuffs (fish and meat, especially), the large number of cancers of the digestive tract observed in the populations of northern Europe, heavy consumers of smoked products, furnish indisputable epidemiological evidence of the dangers they pose to health. The same thing may be said of salting, which increases the risk of hypertension and stroke, and indeed of every method for preserving foods that yields alarmingly high levels of fats and sugars. The truth of the matter is that no wholly safe method exists and never will exist. Sensible nutritionists recognize this. Their advice recalls Paracelsus's wise dictum of almost six hundred years ago: "Nothing is poison. Everything is poison. The difference is in the dose." Long before Paracelsus, of course, the Greeks had stigmatized excess, advocating moderation in all things.

Insincerity founded on food neophobia is therefore not a satisfactory reason for refusing to take note-by-note cooking seriously. But why should we give up traditional cooking and adopt note-by-note cooking in its stead? Or, if this is a false choice, why shouldn't we follow the example of molecular cooking and practice note-by-note cooking alongside traditional cooking? Or combine the two with a view to inventing hybrid dishes? It is the responsibility of proponents of note-by-note cooking to persuade doubters, not the other way around. Let me therefore proceed to consider the most serious objections—without fear of answering them as frankly as possible, for it is my intention neither to sell products nor to propagate an ideology nor to wield power. Quite to the contrary, my purpose is to advance the cause of enlightened connoisseurship. I begin by briefly introducing six sets of questions that will be treated in fuller detail later on, involving technology, nutrition, toxicology, art, economics, and politics.

TECHNOLOGY

A first group of questions concerns the nature of the compounds to be used in note-by-note cooking. Food manufacturers already use very pure compounds, such as water, sodium chloride (salt, in its purest form), sucrose (table sugar), gelatin, and so on. Many people are unaware, perhaps willfully, that these ingredients have long been used in a great many products, whether they are extracted, purified, or modified in various ways. Think of flour and other powdered products, to which anticaking agents are added to prevent them from forming lumps.

A broad range of other compounds might also be used, from saccharides to amino acids, and all the more easily as the food industry already manufactures them. Makers of food additives, for example, market coloring agents, vitamins, and preservatives in addition to polysaccharide gels and other thickening agents. It is true that such additives are not regulated in the same way as food products themselves. But that may change in the not so distant future.

Then there is the question of how pure compounds should be in order to be approved for commercial use. If the standard required by law is not unreasonably stringent, cooks, like the musicians who pioneered the use of synthesizers, will be able to enlarge the palette of ingredients to include simple mixtures—something the food industry already makes, particularly through the milling of wheat and the processing of milk: milling wheat gives us the husk, starch, gluten, and so on; processing milk gives us fats, powders, and various proteinacious preparations. Gelatin, for example, is not pure, in the technical sense that it does not consist of a single sort of molecule, for the extraction of collagen from animal tissues produces a massive dispersion of polypeptide chains. A sheet of gelatin, in other words, contains more or less large molecules of various kinds. Similarly, the starchy matter of vegetables (also known as native starch) is not pure because it is made of two major compounds, amylose (one really ought to say "amyloses" since here again there is no perfect molecular homogeneity) and amylopectin (or amylopectins, for the same reason). Keep in mind, by the way, that since glutinous rice starch is made exclusively of amylopectins, the modified (or processed) starch that is obtained from it is among the products that can be used in note-by-note cooking right away, without any further legal determination regarding admissible degrees of purity. Pastry chefs have been doing something very similar to note-by-note cooking for quite a while now.

Let us return for a moment to *fractionation*, the chemical term for the procedure of separating a mixture into its component parts, which I mentioned a moment ago in connection with vegetable and animal products. From wheat, a variety of ingredients are routinely recovered, including polysaccharides, proteins and amino acids, tensioactive agents; from milk, we get amino acids, peptides, proteins, glycerides, and so on. Why shouldn't the same thing be done using other such products? In principle, laboratory separation processes (filtration, direct or reverse osmosis, cryoconcentration, vacuum distillation, and the like) could be employed to obtain reasonably pure fractions for use in note-by-note cooking.

Government-sponsored research teams in France are hard at work investigating this idea. In Montpellier, for example, experiments are being conducted with specially modified membrane filters to recover water, sugar, acids, and what is known as the "total phenolics fraction" (a brightly colored powder with a powerful flavor) from the juice of grapes. The really interesting thing is that the character of this fraction is strikingly different depending on whether it is extracted from the juice of Syrah or Pinot grapes, for example. The distinctive nature of the initial ingredients, in other words, is not flattened out by the separation process any more than the specific qualities of a piece of meat are masked or diminished by cooking. This means that note-by-note cooking can preserve traces of *terroir*!

Now that we have some idea of the sort of ingredients that can be used, the question arises of how they are to be combined. We mustn't lose sight of two things: first, that most of the foods we eat today are made up mostly of water, and, second, that the solubility of many compounds is reduced in an aqueous environment. This is why emulsification—the dispersion of oil droplets in water—is the primary operation in creating note-by-note dishes. But it is not the only one. Indeed, all techniques of dispersion will have their use.

The various biological properties of foods will need to be taken into account when we look to combine compounds. To be sure, nutritional content is important. But it would be a mistake to forget that if we are to find foods agreeable to eat, they must stimulate various kinds of sensory receptor: visual, olfactory, sapictive (the term I suggest we use in speaking of taste), trigeminal (referring to fresh, spicy, and tingling sensations), tactile, and thermal. This raises a whole host of questions to which we do not yet have any answers. Even if it were possible to determine the light-absorption spectrum of a mixture of pigments whose individual spectra are known, for example, we would not be able to predict the exact color of the mixture. Similarly, mixing odorant compounds in proportions close to their perception thresholds will produce unpredictable results. Not even the result of combining only two odorant compounds is known in advance. Will the mixture have a flavor in which the presence of each of the two compounds can be detected or a new flavor, distinct from that of either compound? No one can say.

Matters are still more uncertain with respect to flavor because the biological receptors of sapid compounds in the mouth remain largely unknown. Receptors of long-chain saturated fatty acids in the papillae were identified fewer than twenty years ago. Their discovery has stimulated a great deal of promising research, much of it yet incomplete. In the meantime, however, there should be no hesitation in using citric, tartaric, malic, acetic, and other acids or saccharides such as glucose and fructose in addition to our old standby, sucrose. With regard to trigeminal effects, a number of refreshing or pungent compounds are known, such as eugenol (noted also for its contribution to the fragrance of cloves), menthol (which has two forms, though only one is perceived to be refreshing, as we shall see later in chapter 3 in connection with flavor), piperine (largely responsible for the pungency of black pepper), capsaicin (a principal source of heat in hot peppers and a variety of spices), ethanol, and sodium bicarbonate.

The consistency of foods is another thriving field of research. Colloidal "soft" materials are poorly understood. Creating multiple emulsions, at first sight a rather straightforward proposition, has turned out to be much more difficult than anyone expected. More generally, anyone who imagines that all the challenges presented by texturing processed foods have been met with the development of surimi and related artificial products is sadly mistaken. Will it be possible one day to fabricate a consistency similar to that of an apple or a pear? So long as it is unclear what making even a prototype artificial fruit would entail, the prospect of large-scale commercial production remains very far off.

In short, there is much to be done. Chefs and food scientists must nevertheless be made to see, as I have already emphasized, that there is little or no reason to reproduce food ingredients that already exist. Just as a synthesizer can copy the sounds of a piano or a violin, note-by-note cooking could copy the flavors and other properties of wines and vegetables and meats—but to what purpose, apart from particular applications, such as dishes famously associated with astronauts working in outer space for long periods of time? We would do better, much better, I believe, to explore that vast continent of dishes that have never yet been created.

A simple calculation will show that the phrase "vast continent," far from being an overstatement, is much more nearly the opposite. Suppose that the number of classical food ingredients is on the order of a thousand and that a typical traditional recipe uses about ten of them. The number of possible combinations is therefore one thousand raised to the power of ten, or 10^{30}. By contrast, if we suppose that the number of different compounds present in foods is likewise on the order of a thousand, but that the number of compounds that will be used in a typical note-by-note dish is on the order of a thousand, rather than ten, then the number of possibilities is roughly 10^{3000}— without taking into account the fact that in note-by-note cooking the concentration of each compound may vary, which means that the new continent is, for all practical purposes, infinitely more immense than the old one. What would be the point, then, of trying to replicate the almost insignificantly small world we already know?

NUTRITION

I need hardly point out that traditional foods are no guarantee of good health. The proof is the pandemic of obesity we presently see in much of the world. No doubt the unbalanced quality of modern diets has not helped matters. But it is probably truer to say that the sheer abundance of food today, unprecedented in the history of our species, has put us in a position for which we were not prepared by biological evolution.

Over the course of many thousands of years humanity has had to confront alternating periods of abundance and scarcity. Nutrigenetics, a branch of nutritional genomics, continues to make progress in identifying the physiological mechanisms that have assured the survival and propagation of the human race. In the past, excessive food intake enabled the body to stockpile fats in anticipation of future shortages, while restricted intake and the gradual loss of appetite it brought about made it easier to adapt to periods of famine as well. The nutritional implications of note-by-note cooking need to be considered in the context of the food industry's use of sweeteners for several decades now and a growing consumer preference for "lighter" foods and beverages (lighter, that is, owing to the addition of air and water), which raise the question of whether low-calorie products cause compensatory overeating. Research now being carried out in this connection will be a fruitful point of departure for the study of the long-term benefits of note-by-note cooking by comparison with the manifest inadequacies of the present regime.

There is also the question of the effectiveness of dietary supplements—vitamins, oligoelements, and so forth—and, indeed, of nutriments in general. Scientists have diligently investigated the various claims made on behalf of such products. Once again, however, it would be naive to believe that all outstanding issues have been resolved. A planned European study of vitamin E supplements (a term that designates a group of hydrophobic compounds having particular antioxidant properties), for example, was recently cancelled because of the abnormally high number of fatalities due to lung cancer and coronary disease among subjects in the supplemented group (many of them, of course, smokers). So far there is no evidence that vitamin E supplements lower rates of lung cancer and coronary disease, but in the absence of further research no firmer conclusion can be reached.

TOXICOLOGY

It is quite true that the harmful effects of some compounds on the human organism are not yet well understood. Extraordinary discoveries continue to be made with encouraging regularity, however, such as the polymorphism of P450 cytochrome enzymes (it turns out that we are equipped with much more elaborate detoxification systems than was previously thought) and, quite recently, the transfer of bacterial genes that colonize seaweed to the intestinal bacteria of persons who eat it.

Yet uncertainty surrounding the long-term epidemiological consequences of note-by-note cooking does not put it in a fundamentally different position than traditional cooking, which uses vegetable and animal ingredients whose innocuousness has never been satisfactorily established. It is one of the paradoxes of modern diet that new foods are subject to far more severe scrutiny than older foods, the sale of many of which would be prohibited if they did not enjoy the advantage of having long been familiar. But note-by-note cooking will be able to avoid the carcinogenic danger of benzopyrenes in smoked products, for example, simply by not using such compounds. In the same way it will be able to avoid the toxic risks associated with myristicin (found in nutmeg), estragole (in tarragon and basil), glycoalkaloids (in potatoes and tomatoes), certain glycosinolates (in cauliflower), and certain phenolic compounds (in various vegetable tissues).

Restrictions on the sale of compounds are likely to resemble the rules that presently govern the sale of liquid nitrogen and ultrasound probes for culinary purposes as well as of the heating elements used to ensure the even distribution of heat in temperature-controlled water baths required by *sous-vide* cooking. The ongoing refinement of cooking techniques will make new forms of regulation inescapable, just as the introduction of gas and of electricity in homes and businesses a century ago made it necessary to take special safety precautions.

Accidents are no doubt to be expected—not because note-by-note cooking is more dangerous than using a kitchen knife, but because the culinary world is no less likely than any other to have its share of negligent or reckless behavior. In July 2009, to mention only the most recent sensational example, a young German cook blew off one of his hands and suffered serious injuries to his lower abdomen and legs when liquid nitrogen he had kept hermetically sealed in a thermos bottle exploded.

ART

Art is obviously a complicated subject. For the moment, I shall say simply that culinary art, like painting, music, sculpture, and the other arts, seeks to arouse our emotions. The culinary artist's fondest hope, after all, is that his guests will look at one another and exclaim, "Oh, that's *good!*"

Just as artists in other fields continually introduce novel elements of various kinds in their works, so too chefs constantly seek to create original sensations. In this respect, at least, note-by-note cooking cannot help but meet with their approval, and that of their customers, because the new possibilities it offers are virtually unlimited.

Producing the first works of note-by-note cuisine nevertheless proved to be difficult. Cooks who accepted the challenge had to learn the chemical alphabet of compounds before they were able to form meaningful gustatory words. Note that I use the past tense here because note-by-note cooking is already a reality. In 2006 I prevailed upon my friend Pierre Gagnaire (who has restaurants in Paris, London, Tokyo, Dubai, Hong Kong, Moscow, Courchevel, Berlin, Las Vegas, and Seoul) to become the first chef in history to produce an entirely note-by-note dish. After several months of work, during which I offered advice, assistance, and encouragement, he presented the result as part of a special dinner in Hong Kong on April 24, 2009. This dish was called "Note-by-Note No. 1" (see the color illustrations).

Then, in the summer of 2010, the Alsatian chefs Hubert Maetz and Aline Kuentz created a note-by-note dish of their own on the occasion of a Franco-Japanese scientific conference in Strasbourg. More recently chef-instructors at Le Cordon Bleu, the culinary arts school in Paris, prepared a full menu of note-by-note dishes for a limited number of guests. And on January 26, 2011, to mark the advent of the International Year of Chemistry, sponsored by UNESCO, Jean-Pierre Biffi and his team at the Paris catering company Potel & Chabot prepared a note-by-note meal for almost a hundred persons. Not a day goes by without some new advance being announced. The reason for this could not be more plain: the true culinary artists of our day are fascinated by a new method of cooking that permits them an unprecedented freedom of expression.

A word for those who fear that this new method spells the end for their beloved pot-au-feu, cassoulet, and choucroute: in the domain of art, there is no such thing as replacement, only addition. This has the consequence that the range of choice is perpetually being enlarged. Debussy did not cause Mozart or Bach to disappear, any more than Picasso and Buffet prevented us from continuing to admire Rembrandt and Brueghel. Similarly, molecular cooking has not done away with nouvelle cuisine or with the style of mixing culinary influences from various cultures known as fusion or with traditional cooking in either its classical or regional forms. Note-by-note cooking will not be any different in this regard.

ECONOMICS

Will note-by-note cooking be more expensive than current methods of cooking? Not only must the cost of investment in new equipment be taken into account, but also the increasing cost of energy. The likelihood that the price of conventional fuels will continue to rise may eventually be decisive in assuring the success of note-by-note cooking. Wasting up to 80 percent of the energy used to heat pots and pans on kitchen stoves is considered acceptable today. It will not be considered acceptable tomorrow, when fossil fuel reserves will be nearly exhausted and energy will have become prohibitively expensive.

What advantage does note-by-note cooking present from this point of view? Consider, for example, the reduction of wine in preparing a sauce. This is mainly a matter of evaporating water. Assuming a reduction of the sort performed by professional cooks (roughly by two-thirds), the energy consumption looks to be on the order of 0.417 kilowatt hours, and the cost about €0.05 (not quite seven cents on a U.S. dollar) per individual serving. That may not seem like much—but wait a while longer. As the price of natural gas moves ever more sharply upward, energy costs will soon become a rather greater source of concern.

The promise of note-by-note cooking can hardly be understated. Dissolve some phenolic compounds extracted from grape juice or wine, together with a little tartaric acid, a little glucose, and a little salt, then heat the mixture for a few seconds until it is warm enough to serve, and you will have consumed virtually no energy (in addition to saving a great deal of time). Obviously some energy will be needed to prepare the ingredients, but if you use the kind of filtration processes that bottled-water companies have been using for several decades now, energy consumption will be much lower than in traditional cooking. What's more, restaurants and caterers will be able to exploit economies of scale: the larger the quantity of food being prepared, the greater the savings. As more and more chefs adopt the techniques of note-by-note cooking, the cost of producing meals in this way will begin to fall and profits will rise, slowly at first, but then at a gathering pace.

The making of sauces is only one example, selected at random from among a thousand or more such examples. Energy consumption has never much mattered in traditional cooking. Still today, no chef thinks twice about cooking meats at temperatures of 200°C (almost 400°F) or more in order to produce compounds that could be obtained much more quickly using the techniques of note-by-note cooking, even in the case of large-scale production, and at a far lower cost per serving. In an oven, it costs just as much to roast one chicken as ten!

Note, too, that not all the compounds used in note-by-note cooking will have to be synthesized. Indeed, it will very often be preferable to extract them from vegetable products. Chemists well remember the years of hard work it took to synthesize vitamin B12. In the absence of a proven experimental method, then, we should look to the farm rather than to the laboratory because obtaining compounds from plants will be both less expensive and more expeditious.

POLITICS

From its inception, note-by-note cooking could not help but stimulate fears that it would force people to eat "chemistry." It revived the dread of an earlier age, when science fiction seemed to herald a world of nutrient pills, soylent steaks, and foods made from petroleum and coal. Here again, as in the case of genetically modified organisms (GMOs), reasoned political debate is contaminated and confused by ideology. The value of note-by-note cooking will become apparent only if uninformed prejudice is combated and persuasive reasons are given for accepting new foods. One of the first to appreciate the importance of the role played by elite tastes in this regard was Antoine-Augustin Parmentier (1737–1813), who persuaded the king of France to eat potatoes, thus setting an example for the people of his kingdom.

Quite apart from the difficulties involved in introducing note-by-note cooking to the widest possible audience, however, ought we not to be worried, as in the case of GMOs, that it will have adverse consequences for traditional forms of social and political organization? What will become of farmers, for example, in the event, however improbable, that all cooking will one day be note-by-note cooking? No one can say with certainty, of course. But just as the owners of vineyards now make more money producing wine rather than grapes, so farmers of all kinds (and not merely large agribusinesses) may be expected to do very well by producing fractions from vegetable and animal products, which today cannot always be sold at prices high enough to be profitable. Note-by-note cooking stands to make farmers better off by encouraging them to manufacture higher-value fractions at the farm for culinary use. Instead of selling fruits and vegetables that are liable to spoil on the way to market, they could sell the extractable parts of these fruits and vegetables. Note, too, that this is where politics links up with the energy question. Transporting fresh fruits and vegetables is mainly a matter of transporting water. Wouldn't it make more sense to eliminate the water at the point of origin and ship lighter products at lower cost?

There remains, finally, the challenge that note-by-note cooking poses for science itself, which has often developed in response to advances in what used to be called the chemical arts. Yet another such occasion presents itself today.

LEARNING TO COOK NOTE BY NOTE

Note-by-note cooking is something that must be learned. In one sense the present situation is no different than the one thirty years ago, when cooks had to be trained in a new style of cooking. But learning note-by-note cooking will be more difficult than molecular cooking because this time a whole new way of thinking about flavor has to be mastered before dishes can be made. With molecular cooking, the tools were new but the ingredients weren't. Cooks were still working with veal, crab, and leeks. They knew what these things tasted like, and they prepared them in more or less familiar ways. With note-by-note cooking, the ingredients themselves are new. Many of them will be unknown to begin with, and cooks will have to learn to combine compounds having unfamiliar properties whose effect on the sensory faculties is apt to be quite different than what we are used to. Thus, for example, beta-carotene, a marvelous powder that even in very small amounts imparts a bright orange to certain substances, has neither flavor nor smell, only color. Salt (sodium chloride) has flavor, but neither smell nor color. Many odorant molecules have no flavor, either because they are used in very weak concentrations or because they are not soluble in water, which constitutes the main part of saliva.

Even so, note-by-note cooking will not seem terribly difficult once cooks have become acquainted with the new ingredients and new methods it involves. When you learn a new language, you need to learn words and rules for combining them to form sentences. In the pages that follow, we shall see how to do the same thing in the kitchen—and have fun doing it.

There's no need to be gloomy in order to be serious.

FURTHER READING

Kurti, Nicholas, and Hervé This. "Chemistry and Physics in the Kitchen." *Scientific American* 270, no. 4 (1994): 44–50.

This, Hervé. *Building a Meal: From Molecular Gastronomy to Culinary Constructivism.* Translated by M. B. DeBevoise. New York: Columbia University Press, 2009.

——. "Molecular Gastronomy: A Chemical Look to Cooking." *Accounts of Chemical Research* 42, no. 5 (2009): 575–583.

——. *Molecular Gastronomy: Exploring the Science of Flavor.* Translated by M. B. DeBevoise. New York: Columbia University Press, 2007.

——. "Molecular Gastronomy Is a Scientific Discipline, and Note-by-Note Cuisine Is the Next Culinary Trend." *Flavour* 2, no. 1 (2013). http://www.flavourjournal.com/content/2/1/1.

——. "De quelles connaissances manquons-nous pour la cuisine note à note?" *L'Actualité chimique* 350 (March 2011): 5–9.

This, Hervé, and Pierre Gagnaire. *Cooking: The Quintessential Art.* Translated by M. B. DeBevoise. Berkeley: University of California Press, 2008.

ONE SHAPE

HOW SHOULD WE GO about creating a note-by-note dish? In creating a traditional dish, some chefs begin by making sketches; others go to the market; others simply sit down and think. Any of these methods, even the second one, might work in note-by-note cooking. But let's try sitting down—in front of an empty plate.

A plate? It may seem an oddly conventional choice considering that what we are trying to do is to imagine a truly new way of cooking. Foods are physical objects, and so they can be served in many different ways. They can, of course, be served in bowls or in cups. But they could also be suspended. Fruits, after all, hang from trees. Why couldn't they be thrown? It was thus, after all, that God caused manna to fall from heaven. In principle, at least, there are any number of ways in which dishes can be presented. But let's not complicate matters too quickly. Let's start with the traditional plate and think about what we wish to put on it.

Gases, vapors, and fumes are possible candidates, at least theoretically, but the density of these substances is so slight that most gourmets will prefer liquids and solids. In the case of a liquid, the question of shape and form arises only indirectly because it is bound to assume the form of the container in which we choose to serve it. But if a food is solid, it will have the shape that

1.1 *AN EMPTY PLATE.*

we ourselves will have given it. In either case, we face the problem of decid-
ing which shapes are most appropriate.

Unless you are a great artist—someone who, almost by definition, cannot
help but surprise his audience—it would be mere laziness, if not actually
childishness, to serve a dish consisting of a single object and a single shape.
Unless times are very hard indeed, a breast of chicken served without sauce
and something to go with it, a vegetable or a starch, is a sorry sight. A tomato
with just a sprinkling of salt likewise makes a poor meal. We may well appreciate its virtues when we pause during a long day's hiking in the mountains
for something to eat, hunger having become the best of cooks, but no civilization in normal times has willingly resigned itself to so little. In both these
examples, as in any number of others that may easily be imagined, the most
disconcerting thing is the lack of *structure*. Cooking might almost, in fact,
be said to be entirely a matter of structure, in the sense that a cook whose
ambition goes beyond executing a series of familiar procedures known as a

"recipe" has no choice but to construct—to use his own imagination in giving a certain structure or shape or form to his dish. He will be aided in this task if he is aware that the human brain has developed the ability over millions of years of evolution to recognize contrasts and differences. When you enter a room filled with cigarette smoke, for example, you are struck at once by the odor of burned tobacco. But if you remain in a smoky room long enough, you no longer perceive the odor. The same is true for color. We are sometimes surprised when, developing film in a darkroom that is artificially illuminated by red light, everything looks red at first, then normal. The reason is that after prolonged exposure to light in which the wavelength that the eye perceives as red is predominant, we no longer see the color red. The same goes for the perception of consistencies, and also of shapes, which are what we are concerned with here.

The question of structure seldom arises in traditional cooking. Shapes are given for the most part by the ingredients themselves, and even if the cook slightly alters them by cutting and chopping and so forth, he is generally limited to assembling a series of shapes given in advance. Why have cooks been content historically to work with so few shapes? Probably because of a practical concern with wasting as little food as possible. A fish filet trimmed in the form of a slab would have been an unthinkable luxury in times of famine, when not a scrap of food was thrown away. Cooking as we know it today developed when shortages were common.

The traditional cook who nevertheless insists on creating novel forms offers proof of the affection he feels for his guests. Cutting a food into a square is a way not only of showing that he is willing to sacrifice a part of the ingredients he has purchased at more or less great cost, but also of showing that he has applied himself to the task of pleasing those who dine at his table. Thus one trusses a roast with butcher's twine to give it a cylindrical shape or slices (or juliennes or dices) carrots instead of leaving them whole. In the case of Pommes de terre à l'anglaise, lopping off the two ends of the potato to make a log and then cutting seven sides or faces to give it the shape of a small barrel is a way of reconciling labor with economy: with seven sides, the cook has to go only to a very small amount of trouble in order to keep the loss of edible material to a minimum. Traditional cooking had the right idea. But it was handicapped by a certain conceptual stinginess, a certain lack of imagination, and failed to get very far.

The problem of structure assumes a different form in note-by-note cooking, but it is no less unavoidable: our empty plate will remain empty if we are unable to give a shape to the foods we wish to construct. How many shapes are there for us to choose from? A much larger number, as it turns out, than traditional cooking would lead us to believe. As a practical matter, the answer to our question is really quite simple: one takes a mold, which already has a shape, and fills it with a substance that will then take on the shape of the mold and retain it. From a theoretical point of view the answer is simple as well, but immensely more illuminating. Since most complex shapes can be reduced to elementary shapes, let's examine this latter class first, beginning with so-called regular forms.

POLYHEDRONS

Polyhedrons are volumes enclosed by flat surfaces. Prehistoric toolmakers, for example, took an irregular piece of flint and modified it by cutting it along a flat plane so that each cut produced a smooth face. A moment's reflection will show that a physical object cannot have only two faces, or only three. The minimum number of faces for a polyhedron is four, in which case the object is a tetrahedron. A tetrahedron can be regular, with faces that are identical, or it can be irregular. The same is true of objects with more than four faces. Furthermore, whether polyhedrons are regular or not, they can be either convex or concave.

CONVEX POLYHEDRONS

In the history of mathematics, regular convex polyhedrons occupy a special place. They were systematically studied by the Greek mathematicians of antiquity, who interpreted them as models of physical objects. The Greeks understood that although it is possible to construct an infinite number of convex polyhedrons, only five of them have faces that are regular polygons, which is to say polygons having equal sides. These objects, the so-called Platonic

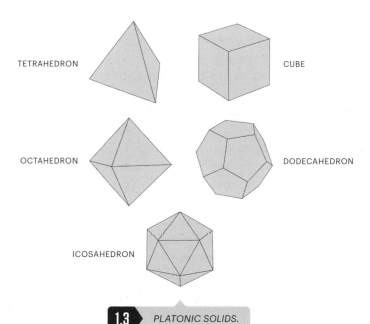

TETRAHEDRON

CUBE

OCTAHEDRON

DODECAHEDRON

ICOSAHEDRON

1.3 *PLATONIC SOLIDS.*

solids, are the tetrahedron (with four triangular faces), the cube (six square faces), the octahedron (eight triangular faces), the dodecahedron (twelve pentagonal faces), and the icosahedron (twenty triangular faces).

Our names for convex polyhedrons of all kinds, not just regular polyhedrons, are derived from Greek roots. I have just mentioned the tetrahedron (from *tetra,* meaning four); beyond this, there are pentahedrons, hexahedrons, heptahedrons, octahedrons, nonahedrons, decahedrons, hendecahedrons, dodecahedrons, and so on. But knowing their names will count for little so long as we remain unable to rapidly survey the universe of possible forms and then, from this very, very long list, select the ones that will fill up a plate in just the way we have in mind. Taking symmetry into account will help us sort through the most promising candidates far more efficiently.

If we look at a person's face, for example, we see that it is symmetrical: a vertical plane passing through the nose divides the face into two roughly identical halves, reversed in relation to the plane. A plane of this sort is called a "plane of symmetry" (or a "symmetry plane"). In other cases, the relevant geometrical perspective is provided not by a plane, but by an axis. Take, for example, a prism with a triangular base—in other words, an object described by three vertical planes passing through the sides of a triangle traced on a horizontal supporting plane. If the base triangle is equilateral, with three equal angles, then the vertical axis, which passes through its center, exhibits what geometers call symmetry of order 3: rotating the prism by a third of a turn brings it back into alignment with its initial position. Rotational symmetries of order 4, 5, 6, and so on are defined in a similar fashion.

And so it is that number makes itself manifest in the world! Is it visible to us because it does in fact constitute the basis of the physical world or because our minds have been predisposed by biological evolution to find it everywhere? This is a very difficult question—but cooks, even without trying to answer it, will be able to exploit the mathematical properties (and, more generally, the cultural associations) of numbers in the kitchen. Starting from 4, the minimal number of faces of a polyhedron, we move on to 5, which is sometimes thought of as a magic number, and then to 6, whose ambivalent reputation in the popular mind is due to its role in games of chance and prophecy. I mentioned the number 7 a moment ago in connection with the seven faces of Pommes de terre à l'anglaise. Whether this number somehow minimizes both the loss of edible material and the labor required to make certain dishes, as legend would have it, it has special significance for many people: the seven basic colors, the seven days of the world's creation, the seven pillars of wisdom, the seven days of the week, and so on. The number 7 is a mystical number for some, the symbol of perfection; and for at least one fraternal organization that comes to mind it holds greater importance than perhaps it should. Seven is also the number of chapters in this edition of my book.

From 7 we go on to 8, a symbol of happiness responsible for the congestion in town halls on the eighth day of the month in France and other countries, when many couples come to be married. The number 9 is considered to be a token of good luck, whereas 11 is linked with the idea of excess. As for 13, there is no need for any very lengthy explanation. In our culture it has a baleful connotation, and cooks, for their part, will generally want to avoid it; but if they happen to be cooking for a party of guests for whom 13 is a propitious number, there is no reason not to use it in selecting the shapes of various dishes. Is this mere opportunism? No doubt it is, at least to some extent. Still, even if the cook is imagined to be a simple, naive soul who is in the habit of choosing the shapes of the food he makes on the basis of his own numerological whims and prejudices, what harm is there in his trying from time to time to add to the happiness of the people he feeds by catering to theirs?

CONVEXITY AND CONCAVITY

This exceedingly rapid tour of polyhedrons would not be complete without saying a word or two about convexity and concavity. Why are some objects pointed or protruding, and others hollow? One hardly needs to be preoccupied by sex in order to think in the first place of the complementarity of masculine and feminine. But there is also the relation between container and contained. Thus a concave polyhedron can accommodate either a liquid or a solid substance. What is more, a hollow object that is open on one side can be sealed or closed up, which suggests to the mind of a cook the idea of stuffing foods inside one another—a way of surprising the person to whom they are served.

Which form should we prefer, concave or convex? Consider the difference between the English place setting, where the fork is placed with its tines upward, and the French place setting, where the fork is placed with its tines downward. The choice in this case conceals a question of politeness having to do with one person's sympathetic concern for the psychological and physical comfort of others: a fork placed in the English manner confronts the person across the table with a set of sharp, menacing prongs, whereas the same fork, turned face down, presents a softer, more rounded aspect. Again, as a matter of courtesy, the blade of the knife is customarily turned toward the plate, and not outward in the direction of the person sitting next to us.

Showing concern for those who dine at our table is not merely a detail of etiquette, however. It is a form of conviviality in the literal sense of the term. From the geometrical point of view, the cook must take care not to take too pointed an interest, shall we say, in the sensibilities of his guests. Apices and convex polyhedrons should not be banned from the repertoire of admissible culinary forms, of course—but only on the condition that all the rough edges have been smoothed out!

NONPOLYHEDRAL SOLIDS

Another, more rounded class of shapes that recommends itself to the cook's consideration includes the sphere, the cylinder with a circular base, the cone. Mathematicians have exhaustively studied all of these solids. In all of them, the idea of symmetry is essential.

ROTATIONAL SYMMETRY

Spheres, cylinders, and cones are members of a family of shapes that exhibit rotational symmetry; that is, they are produced by rotation around an axis, without undergoing any alteration of shape. A sphere is produced by the rotation of a semicircle around an axis passing through the center of this semicircle. A cylinder exhibiting rotational symmetry is produced by turning a rectangle around one of its edges. A cone is likewise the result of rotating a triangle. Other solids can be obtained by rotation as well. Most bottles, for example, exhibit rotational symmetry around their axis. Pottery made on a potter's wheel displays the same property.

All these objects have a pleasing roundness, long appreciated not only by architects and interior designers, but also by sculptors, stoneworkers, and furniture makers. Today more than ever, with the advent of note-by-note cooking, cooks will profit from the example of these and other craftsmen, whose ancestors were not given shapes to work with in advance—the shape of a tenderloin of beef, for example, or a breast of chicken or a filet of fish. They had to learn which shapes were most suitable for their work. Their knowledge and experience can now be imported into cooking.

OTHER FORMS OF SYMMETRY

Shapes produced by rotation around an axis are by no means the only possible symmetrical forms. Many living creatures are more or less symmetrical, but their symmetry is not due to rotation. A fish, for example, is roughly symmetrical in relation to a plane. An idealized starfish, with its five arms, has a symmetry (of order 5) in relation to an axis perpendicular to the plane of the arms. One thinks, too, of star fruit (carambola or Chinese gooseberry) and the inflorescence of a sunflower, both of which exhibit a symmetry that is more difficult to recognize—also of squashes that have several lobes and of winter squash and pumpkins, which have a larger number of lobes more or less evenly distributed around an axis.

A whole book could be written about symmetries—indeed, many such books already have been written. I shall limit myself here to observing that symmetry, in the world in which we actually live, is an unattainable ideal: our faces are not composed of two absolutely symmetrical halves in relation to the nose; more generally, shapes that appear at first sight to be perfectly balanced are marred by a slight departure from symmetry, even if our brains usually manage to cope with the imperfection readily enough. It is, after all, an old painter's trick to introduce a slight asymmetry just where the mind expects to find symmetry. This little blemish is what makes formal perfection—which Nietzsche rightly saw as poisonous and inimical to life—human and loveable. In cooking no less than in many fields of endeavor, not only painting and the other arts but also the sciences, even the worlds of administration and commerce, we are well advised, I believe, to tolerate some small degree of imperfection lest we become inhuman.

WHEN SYMMETRY DISAPPEARS

Cooks should therefore be guided by the simplicity found in symmetry. But they mustn't neglect complexity, because the human mind perpetually hesitates between the two. Nevertheless cooks should not be interested in complexity for its own sake—too much of it and their creations will be incomprehensible. In my book with Pierre Gagnaire, I proposed (unwisely, perhaps, since when it comes to art we should be wary of rules) that art can be appreciated only if it introduces a limited and controlled sort of novelty into a composition whose elements are for the most part recognizable and familiar to us. It will therefore be to the artist's advantage to slightly deform or distort the known and the familiar, whether he is dealing with a geometric shape or an abstract "form."

Why the scare quotes around the last word? Because even though we are concerned primarily with geometry here, it is impossible to resist the temptation to venture a useful generalization involving Platonic forms—the name Plato gave to objects of intelligible (rather than visible) reality, objects of pure knowledge. In this sense, one may speak of the Orangeness of oranges, the Bitterness of beer, the Freshness of mint, the Spiciness of pepper, and so on. These capitalized nouns refer to ideal archetypes, of which the actual objects we know from firsthand experience are regarded as so many inferior copies. I am well aware, of course, that most gourmets will balk at the suggestion that in the realm of odors, for example, there exists an ideal Vanilla odor. They know that no two neighboring pods of vanilla in a market smell the same; indeed, chemical analysis of the constituent compounds confirms that they differ very slightly. We ought, then, to speak instead of the various odors of vanilla, not of Vanilla as an archetypal odor. The same is true for oranges, whose fragrance cannot be reduced to the olfactory effect of a compound called limonene, any more than the smell of Roquefort can be reduced to the effect of heptanone. Or, again, considering the matter from the standpoint of color: the yellow plum known in France by the name "mirabelle" displays a variety of yellows in addition to shades of orange and red. Similarly, with respect to taste, different kinds of beer display different kinds of bitterness (just as their color may vary from blond to dark over a range of shades); with respect to trigeminal sensations, different kinds of mint (peppermint, penny-royal, bergamot, spearmint, wild mint, and so on) display different kinds of freshness, and different kinds of pepper (from Madagascar, Szechuan, Sar-awak, and many other places) display different kinds of spiciness.

Cooks will nonetheless gain new insight if they keep in mind the similari-ties, rather than the differences, between these qualities, for they will then be able to take advantage of the brain's innate tendency to detect general prop-erties—geometric forms, olfactory forms, sapid forms, and so on—in creating new dishes. A gifted cook can season a sauce with tarragon, for example, so very slightly that the herb's aroma is not obvious: the gourmet tastes, smells something, and begins to wonder. "What's that flavor? Could it be [chews pensively], or perhaps [thinks a bit more]...?" And then suddenly it hits him: "Tarragon!" Surely the same effect could be exploited in the case of both geo-metric shapes and abstract forms, but only so long as cooks are acquainted with all those solids that do not display symmetry—much more numerous, as it happens, than the ones that do.

How are we to begin hacking our way through this jungle? By putting the resources of topology to work. Topology is a branch of mathematics that con-templates objects as though they were made of modeling clay—as though they were capable, in other words, of being deformed, though not to the point of being torn. From this point of view, a ball and a cube are topologically equiv-alent. Think of the ball as an egg yolk that has been cooked at 67°C (153°F) until it assumes the consistency of an ointment, so that it is malleable while yet conserving the color and taste of a fresh egg, and you will see at once that it can be made into a cube.

Topology classifies objects by the number of holes they have. A coffee cup, for example, is topologically equivalent to a doughnut or a rubber tire or a simple knot: all of these things have one hole, as you can easily see by play-ing with modeling clay yourself. Eyeglasses have two holes before they have been fitted with lenses, however, and so are not equivalent to coffee cups.

1.4 TOPOLOGICALLY, A DOUGHNUT AND A CUP ARE IDENTICAL.
(1.4A © ANGEL SIMON-FOTOLIA.COM.)

None of this tells us which shapes we should choose, of course. But it does help us to think more clearly and to get our bearings in an infinite universe of possibilities.

The problem of choosing culinary forms will become somewhat more tractable if we remember that painting, sculpture, music, and the other arts have often imitated nature. Why they have imitated nature and whether imitating nature is a good thing are questions that do not concern us for the moment. What matters is that artists have seldom tried to reproduce nature exactly. Instead they have sought to give both a personal vision and, even more important, a cultural interpretation of it. In the Middle Ages, for example, rather than make objects and figures in the foreground larger than the ones in the background, painters scaled them in a way that corresponded to their cultural significance or social prominence.

Did the formulation of rules of perspective in Florence in the fifteenth century represent a step forward? Certainly they provided painting with a technique that until then had been beyond its reach and that enabled it to draw ever nearer to the precision that photography was later to achieve. But should art concern itself mainly with exact reproduction? For that purpose, surely, reality itself will do. If art seeks above all to arouse emotion, however, the rules of linear perspective represent a step backward, one that Picasso and other painters of his generation adamantly refused to take.

In the art of cooking, social ties and the personal experience of sharing count for more than anything else. This is why, ultimately, I prefer to examine the problem of culinary structure in terms of purpose or intent. What is the cook trying to do? Many things—but in the first place he or she is trying to fill a plate with shapes that will provoke an emotional response from his guests. How shall the culinary artist decide which shapes are best suited to this purpose? This is the basic question that we must deal with at the very outset. It goes to the heart of the mystery of artistic creation, which may be expressed in the form of another question: How is it that a great artist's work comes to be admired by people in every country and in every age despite the mutable nature of personal tastes?

THE FABLE OF THE MAN WITH THE GOLDEN BRAIN

A poetic answer to this second, still more fundamental question is given in a fable by Alphonse Daudet (written in collaboration with Paul Arène) that first appeared in his collection *Lettres de mon moulin* (*Letters from My Mill*, 1869). To "the Lady who asks for light-hearted stories," Daudet replies with these words:

> On reading your letter, madame, I had a feeling of remorse. I was angry with myself for the rather too doleful color of my stories, and I am determined to offer you today something joyous, yes, wildly joyous.
>
> For why should I be melancholy, after all? I live a thousand leagues from Parisian fogs, on a hill bathed in light, in the land of tambourines and Muscat wine. About me everything is sunshine and music; I have orchestras of finches, choruses of tomtits; in the morning the curlews say: "Cureli! cureli!"; at noon, the grasshoppers; and then, the shepherds playing their fifes, and the lovely dark-faced girls whom I hear laughing among the vines. In truth, the spot is ill-chosen to paint in black; I ought rather to send to the ladies rose-colored poems and baskets full of love-tales.

But no! I am still too near Paris. Every day, even among my pines, the capital splashes me with its melancholy. At the very hour that I write these lines, I learn of the wretched death of Charles Barbara, and my mill is in total mourning. Adieu, curlews and grasshoppers! I have now no heart for gayety. And that is why, madame, instead of the pretty, jesting story that I had determined to tell you, you will have again today a melancholy legend.

Once upon a time there was man who had a golden brain; yes, madame, a golden brain. When he came into the world, the doctors thought that the child would not live, his head was so heavy and his brain so immeasurably large. He did live, however, and grew in the sunlight like a fine olive tree; but his great head always led him astray, and it was heartrending to see him collide with the furniture as he walked. Often he fell. One day he rolled down a flight of stairs and struck his forehead against a marble step, upon which his skull rang like a metal bar. They thought that he was dead; but on lifting him up, they found only a slight wound, with two or three drops of gold among his fair hair. Thus it was that his parents first learned that the child had a golden brain.

The thing was kept secret; the poor little fellow himself suspected nothing. From time to time he asked why they no longer allowed him to run about in front of the gate, with the children in the street.

"Because they would steal you, my lovely treasure!" his mother replied.

Thereupon the little fellow was terribly afraid of being stolen; he went back to his lonely play, without a word, and stumbled heavily from one room to another.

Not until he was eighteen years old did his parents disclose to him the monstrous gift that he owed to destiny; and as they had educated and supported him until then, they asked him, in return, for a little of his gold. The child did not hesitate; on the instant—how? by what means? the legend does not say—he tore from his brain a piece of solid gold as big as a nut, and proudly tossed it upon his mother's knees. Then, dazzled by the wealth that he bore in his head, mad with desires, drunken with his power, he left his father's house and went out into the world, squandering his treasure.

From the pace at which he lived, like a prince, sowing gold without counting, one would have said that his brain was inexhaustible. It did become exhausted, however, and little by little one could see his eyes grow dull, his cheeks become more and more hollow. At last, one morning, after a wild debauch, the unfortunate fellow, alone among the remnants of the feast and the paling candles, was alarmed at the enormous breach he had already made in his ingot; it was high time to stop.

Thenceforth he led a new kind of life. The man with the golden brain went off to live apart, working with his hands, suspicious and timid as a miser, shunning temptations, trying to forget, himself, that fatal wealth which he was determined never to touch again. Unfortunately, a friend followed him into the solitude and that friend knew his secret.

One night the poor man was awakened with a start by a pain in his head, a frightful pain. He sprang out of bed in deadly alarm, and saw by the moonlight his friend running away, with something hidden under his cloak. Another piece of his brain stolen from him!

Some time afterward, the man with the golden brain fell in love, and then it was all over. He loved with his whole heart a little fair-haired woman, who loved him well too, but who preferred her ribbons and her white feather and the pretty little bronze tassels tapping the sides of her boots.

In the hands of that dainty creature, half bird and half doll, the gold pieces melted merrily away. She had every sort of caprice; and he could never say no; indeed, for fear of causing her pain, he concealed from her to the end the sad secret of his fortune.

"We must be very rich," she would say.

And the poor man would answer:

"Oh, yes! very rich!" and he would smile fondly at the little bluebird that was innocently consuming his brain. Sometimes, however, fear seized him; and he longed again to be a miser; but then the little woman would come hopping towards him and say:

"Come, my husband, you are so rich, buy me something very costly."

And he would buy something very costly.

This state of affairs lasted two years; then, one morning, the little woman died, no one knew why, like a bird. The treasure was almost exhausted; with what remained the widower provided a grand funeral for his dear dead wife. Bells clanging, heavy coaches draped in black, plumed horses, silver tears on the velvet—nothing seemed too fine to him. What mattered his gold to him now? He gave some to the church, to the bearers, to the women who sold immortelles; he gave it on all sides, without bargaining. So that, when he left the cemetery, almost nothing was left of that marvelous brain, save a few tiny pieces on the walls of his skull.

The people saw him wandering through the streets, with a wild expression, his hands before him, stumbling like a drunken man. At night at the hour when the shops were lighted, he halted in front of a large shop window in which a bewildering mass of fabrics and jewels glittered in the light, and he stood there a long while gazing at two blue satin boots bordered with swan's-down. "I know someone to whom those boots would give great pleasure," he said to himself with a smile; and, already forgetting that the little woman was dead, he went in to buy them.

From the back of her shop the dealer heard a loud outcry; she ran to the spot, and recoiled in terror at sight of a man leaning against the window and gazing at her sorrowfully with a dazed look. He held in one hand the blue boots trimmed with swan's-down, and held out to her the other hand, bleeding, with scrapings of gold on the ends of the nails.

Such, madame, is the legend of the man with the golden brain.

Although it has the aspect of a fanciful tale, it is true from beginning to end. There are in this world many poor fellows who are contented to live on their brains, and who pay in refined gold, with their marrow and their substance, for the most trivial things of life. It is to them a pain recurring every day; and then, when they are weary of suffering....

Alas, my dear cooks, it may be thus that you will decide which shapes to assemble on a plate in order to give form to a dish that you yourself will have constructed: by extracting all the gold there is in your brains—and this at the risk of great suffering, even of death. Out of love for those whom you have invited to dine at your table, you dare to exhaust your own most precious resources in seeking to recover nuggets of artistic truth—your truth, the product of your deepest feeling and emotion.

Some may feel that a book that aspires to inaugurate a new way of cooking should be wary of starting out on such a gloomy note. After all, a fable is just that—a fable. Against the dark melancholy of Daudet's windmill, plunged into mourning by Barbara's death, let us therefore set the luminous prospect of the birth of a truly new cuisine. My sisters, my brothers, all of you who are cooks, if you are true artists, your spirit will speak through your creations. And in giving voice to your spirit, you shall have achieved your supreme purpose as a human being, by sharing yourself, your ideas, your passions, your innermost longings with all those who are dear to you.

FURTHER READING

Berggren, Lennart, Jonathan Borwein, and Peter Borwein, eds. *Pi: A Source Book*. 3rd ed. New York: Springer, 2010.

Coxeter, H. M. S. *Regular Polytopes*. 3rd ed. New York: Dover, 1973.

Neugebauer, Otto. *The Exact Sciences in Antiquity*. 2nd ed. New York: Dover, 1969.

This, Hervé, and Pierre Gagnaire. "Thomas Aquinas and the Green of the Grass." In *Cooking: The Quintessential Art*, translated by M. B. DeBevoise, 166–178. Berkeley: University of California Press, 2008.

Zee, A. *Fearful Symmetry: The Search for Beauty in Modern Physics*. 3rd ed. With a foreword by Roger Penrose. Princeton, N.J.: Princeton University Press, 2007.

TWO

CONSISTENCY

IN THE FIRST CHAPTER, where we started from the empty plate that the cook seeks to fill, I barely touched upon the question of consistency, ignoring gases and considering only liquids and solids. But these are crude notions. Cooks know perfectly well how to make things that are much more interesting than ordinary liquids or common solids: emulsions, sauces of various kinds, pastries, and so on. Between solid and liquid an entire continent remains to be explored—an almost infinite number of variations waiting to be discovered by anyone who is willing to go to the trouble of producing them. This is a truly magnificent opportunity, for we have seen that a dish is created by assembling masses having not only different shapes, but also different consistencies. The art of combining them in a harmonious whole accounts for a large part of the happiness we feel when we eat food that has been expertly prepared.

Recall that our senses evolved in such a way as to perceive contrasts. Earlier, in discussing shape and form, I should perhaps have mentioned that David Hubel and Torsten Wiesel were awarded the Nobel Prize in Physiology in 1984 for their work exploring the mechanisms of visual recognition and, in particular, for explaining how the brain is able to recognize edges in images—a topic of more than passing interest to cooks. In the chapters that follow, we shall look more closely at taste, odor, trigeminal sensations, and color. For the moment, however, let us examine the culinary prohibition against serving a dish that has a homogeneous or uniform consistency. The reason for this, of course, is that the crispness of the skin of a roasted chicken is indispensable if the tenderness of the meat itself is to be fully appreciated; so, too, carving a roasted leg of lamb at an angle perpendicular to the bone, in the French manner, presents a whole palette of contrasting colors and textures, from the crispy brown of the exterior to the bloody red of the meat nearest the bone. But there is much more to the matter than meets the eye.

A WOEFUL MISUNDERSTANDING

It is perhaps not surprising that it is mainly scientists studying the appearance of foods who maintain that this aspect is preponderant in the experience of eating (measurably so, some of them would even argue). Most cooks would say that the odor of a roasted chicken that permeates the kitchen precedes the sight of the crispy golden skin and that we salivate when we smell the odor of baked cheese, not when we look at it. This may well be true. But if the color of a wine or the visible texture of a food leads us to anticipate a taste or a flavor that turns out not to be present (or, in the extreme case, makes us believe that something soft is really hard), the confusion lasts only for a few moments. During the course of tasting, whatever false expectations we may have formed on the basis of visual inspection are quickly dispelled as a result of the perception of consistency by the teeth (which sense degrees of pressure) and by the tongue. Together they set us straight by identifying the nature of the object being consumed and, crucially, by releasing sapid and odorant molecules. Indeed, when a dish is really good, one closes one's eyes in sheer delight, dispensing with visual perception altogether and concentrating on exactly that part of the flavor that is devoid of it. Consistency is an essential element of our appreciation of food, as cooks well know. This is why they are so concerned with cooking times: leave a pan on the stove a moment too long, or remove it from the flame a moment too soon, and the dish will have the wrong consistency.

A word or two of explanation may be in order here. Some readers will wonder why I have so far spoken mainly of "consistency" rather than of "texture." The short answer is that these terms refer to different things. Because we modify a food as we eat it, a single consistency can assume a number of different textures. The situation may be likened to what happens when you dive into a swimming pool. If the entry is clean, almost noiseless, you pierce the water, causing it to part smoothly in front of you. But if the dive is misjudged, your body lands on top of so great a surface that the water doesn't have time to part, because water cannot move faster than the speed of sound in water—a bit like an airplane that cannot break the sound barrier. In either case, however, whether the dive is perfectly or disastrously executed, water is water, and its consistency remains unchanged. Yet the same water, which seems liquid in the case of a clean entry, seems hard as cement in the case of a belly flop. We must distinguish, then, between the consistency of water, a characteristic that is independent of our perception of it, and our perception of the water's consistency, which is to say its texture.

The same thing is true of foods. Here again we must distinguish between consistency, which is a consequence of their molecular structure, and what we actually perceive, which is their texture. Mayonnaise, for example, seems thick if it is eaten slowly, whereas it seems fluid if we wolf it down. In this chapter it is consistency that we are concerned with—and, in particular, the question whether consistency is a primary, or instead a secondary, component of the pleasure of eating. Almost thirty-five years ago, in 1979, two American physiologists, H. G. Schutz and O. L. Wahl, published the results of a study purporting to show that consistency is unimportant. Readers should have been skeptical from the first. After all, why should cooks have worked so hard for so many hundreds of years to get the consistency of roasted meats just right? Why should they have taken such great pains to invent and prepare so many subtly different sauces?

Respondents in Schutz and Wahl's study were asked to rank the relative importance of the characteristics of appearance, flavor, and consistency (which they called "texture," but which should, for the reasons I have just given, be called "consistency") for ninety-four products. These products had been chosen with reference to four criteria: they were easily obtainable in supermarkets; they were available in a form that made them easy to consume; they were widely consumed; and they were consumed without any preparation or admixture with other foods. The study was done in and around Sacramento, California, by means of questionnaires sent by mail: after two reminders, 994 of 1,000 questionnaires sent had been returned, but only 420 of them had been correctly filled out. This alone is enough to suggest how doubtful the results really were. Once the results had been tabulated, it became clear that odor and flavor were unambiguously ranked ahead of appearance and consistency: more than nine persons out of ten assigned greater importance to flavor than to any of the other criteria for each of the ninety-four products.

But wasn't this a foregone conclusion? Since we are incapable of discriminating between the various sensations that make up the flavor of the foods we eat, flavor is the only thing we have to go on. A study establishing the preponderance of flavor in gustatory perception is quite ridiculous: it amounts only to demonstrating that flavor is something more than its various components—that the whole is greater than any one of its parts.

The reason we are unable to isolate the various sensations we detect while we are eating is that these sensations interact with one another. A simple experiment will illustrate the point. Put on a headset with earphones and a microphone, then connect the headset to a computer so you can hear the sounds that are transmitted and processed by the volume-control program. Now eat a flaky, freshly baked croissant, for example, adjusting the volume control as you chew: the croissant will seem to have more of a crunch to it the higher the volume, less of a crunch the lower the volume—proof that our perception of a food's consistency, its texture rather than the consistency itself, depends in part on the noise a food makes when it is chewed.

The study by Schutz and Wahl is vulnerable to another objection, that it was uncontrolled by any experiment that might have served as a safeguard against baseless conjecture and reckless misinterpretation. (The wonderful thing about the experimental sciences, by the way, is that they prevent us from claiming whatever we like in the absence of corroborating evidence.) A study of consistency done without any product actually being chewed, and without any qualitative analysis of the results of mastication, can hardly be taken seriously. Worse still, Schutz and Wahl failed to control for a well-known bias in sensory research. Subjects who take part in tasting studies often answer the questions put to them by experimenters in a way that conforms to their own preconceptions. If, for example, they believe that consistency is less important than flavor, they are likely to give more weight to flavor in their responses. But let's face it: the reality is that people do in fact like and buy many products—crackers, pâtés, and so on—for their consistency.

How much importance should be attached to consistency, then? For a long while it was accepted that solid foods act on appetite by means of a sort of digestive conditioning: because they take longer to be digested and absorbed than an equivalent amount of liquid derived from them, we have the impression that they satisfy physiological needs more fully. Experimental tests of this hypothesis were few, however, and most of these suffered from the same methodological defect I just mentioned (technically known as omitted variable bias). A 1978 study comparing an orange and a comparable quantity of orange juice, for example, concluded that the orange was more filling. Alas, the authors failed to take into account the fact that an orange contains forty times more pectin, which reduces gastric evacuation and thereby contributes to the sensation of satiety. To do the experiment properly, they would have had to compare an orange and a quantity of juice having the same pectin content.

A far better-designed study was conducted more recently by Hélène Rosier-Labouré at the European Centre for Taste, Food, and Nutrition Sciences in Dijon. Rosier-Labouré and her team studied the short- and long-term effects in rats of consuming two versions of a soup consisting of meat, beans, cream starch, and water. The soup was presented in two "textures" (more accurately, as I say, consistencies), one a rough mixture of meat and vegetable matter, the other a purée of the same ingredients. Over the course of three weeks the rats consumed three times more of the chunky version than of the puréed version: not only did they eat the first kind more frequently, but their consumption per meal was greater as well.

This short-term result seemed to contradict the idea that foods are perceived to be more filling the more chewing they require. But it might also be the case that chunky soups are more palatable than puréed soups. Because postingestive effects are typically observed that do not appear in short-term studies, however, the experiment was repeated for a period of six weeks, this time comparing the behavior of rats offered only the chunky soup with ones that were offered only the puréed soup or else the soup of their choice. Over this longer term, the rats fed with the puréed soup ate more than the rats fed with the chunky soup, whose intake declined beginning in the third week.

Are these same effects found in humans? In a related set of experiments, Rosier-Labouré studied the feeding patterns of twelve subjects, which seemed to indicate that the hedonic effect (the appeal of the meal in and of itself, which is to say its flavor) was negligible, whereas there were significant differences in the duration of eating, with the chunky soup being chewed for a longer time—logically enough—than the puréed soup. This observation could be interpreted just the opposite way, however. It may be that the longer chewing time corresponded to the detection of greater flavor.

THE RELATION BETWEEN CONSISTENCY AND FLAVOR

All of this is very well and good, you may say, but surely cooks do not need food science and technology to tell them that the flavor of foods depends in part on their consistency—it should be obvious to anyone who knows what these words mean!

Let us therefore start all over again, this time by eating an apple. We discover that a variety of sensations come together: taste, temperature, texture, odor, color, and so on. But while we do indeed perceive each of these things, we find it very hard to isolate them. We register a synthetic, or general, sensation: the apple's flavor (or, more precisely, the flavor of this particular apple). We can try to try to determine if the taste is sweet or sour, for example, but we cannot consider it independently of the odorant compounds that are released by the destructuring of the apple's fleshy tissue when we chew, rising up from the rear of the mouth into the nose via the retronasal passages (the same passages that carry water from the mouth to the nose when accidentally we get a gulpful while swimming). When we eat, in other words, we detect odors; we do not perceive taste alone, we also perceive a food's smell. The same is true for consistency, the perception of which is complicated by taste and smell.

It is therefore a very clever person who can say what the true consistency of a food is. Texture presents no problem: it is whatever we perceive when we bite into something. But consistency is something different. To say that the flavor of foods depends on their consistency is trivial, then, a sort of pleonasm. Nevertheless it is true that meat juices acquire a special flavor when they are thickened to make a sauce. It is true, too, that a sorbet may taste sharp or dull depending on whether it is made with an ice cream maker or with the aid of liquid nitrogen, that a sour mayonnaise loses its harsh edge with additional beating, and that jelly loses its flavor when it becomes sticky—that is, when it has too much pectin or other gelling agent, which is what causes it to harden.

For centuries, mastering consistencies was a matter of trial and error. Traditional cooking long ago discovered that the gelatin extracted by heating bones and meats in water was useful in making stocks that themselves could be reduced further to form glazes. Without knowing it, cooks did three essential things: they dissolved the protein known as collagen, which gives meats their toughness; dissociated the three molecular strands that make up collagen; and then concentrated these strands. In this way it became possible to impart viscosity to liquids, bind odorant molecules together, trigger reactions that produce such molecules, and emulsify the fatty matter (usually butter or cream) used to thicken sauces. If all these wonderful discoveries were able to be made through the patient accumulation of empirical knowledge alone, how many more are there waiting to be made by cooks who have a theoretical understanding of the physicochemical basis of these phenomena?

Pastry chefs, for their part, learned long ago to cook fruits in order to extract pectin, which they then used to make jams and jellies set. But how much better these same jams and jellies were once the underlying chemistry was known! Pectin molecules are present in fruits that have reached a certain stage of ripeness. Along with cellulose molecules and other polysaccharides, they constitute the cell walls of most plants. Heating fruits in liquid causes the pectin molecules to dissolve, forming a hot syrup; as the syrup cools, the molecules extracted in this way become linked together and form a gel so long as the sugar concentration remains within a range of 55 to 75 percent. The practical expertise developed by jam makers over the centuries nevertheless falls short of what we want to know. Modern chemists discovered that different pectins have different gelling properties. It had long been understood, of course, that when jams are made in copper pans, the copper produces

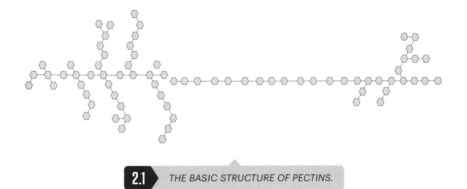

2.1 *THE BASIC STRUCTURE OF PECTINS.*

a firmer gel by binding the pectin molecules together (over time copper becomes toxic, however, and can be substituted for by calcium, which has the additional advantage of strengthening our bones). Commercial research on thickening, gelling, and texturizing additives in recent years has shown that some pectins are esterified, which is to say that they contain methyl groups (groups of one carbon atom and three hydrogen atoms) that modify gelling properties. More remains to be learned about these properties.

But do we have the right, as purists who truly love food, to add pectins to jams? By all means! Cookbooks have long recommended doing just this, particularly in connection with certain kinds of cherry jam: in the event the jam fails to set, the cook is advised to put a muslin or cheesecloth bag containing apple skins and seeds (which have a high pectin content) in the pan. The very same expedient is resorted to today in the commercial manufacture of gelling agents, where the skin extract of apples and citrus fruits is used to produce pectins of various qualities for both home cooks and the food-processing industry.

Pectins and gelatins are not the only possible gelling agents. Manufacturers of emulsifying, thickening, and gelling additives have refined traditional methods for modifying textures, with the result that the dairy industry is now able to make Chantilly creams and other products that withstand the rigors of long-distance transport far better than they did in the past. With the addition of emulsifying compounds to augment the expansive (that is, foaming) properties of cream, and of thickening compounds to stabilize the cream, dairy products that are shaken in the course of handling and shipping no longer separate.

Today the variety of such additives is rather extensive: gum arabic, extracted from the acacia tree; carob gum and guar gum, extracted from seeds (polysianes, which prevent infants from spitting up, are derived from carob gum, used for this purpose since ancient times); fruit pectins; alginates and carrageenans, extracted respectively from brown and red algae; gelatins, extracted from animal bones and hides; various emulsifiers obtained from natural fats by distillation or from natural fats and sugars by chemical reaction, or by extraction from soybeans; xanthan gum, secreted by bacteria; and so on. The opportunity of creating new and varied consistencies that all these compounds present is one that cooking today can no longer afford to neglect, especially in view of the abiding obsession with making foods lighter—chiefly by increasing their water and air content. Nothing, after all, is lighter than air!

The roots of this tendency go back to the early twentieth century, when sauciers began to replace flour, disliked for its blandness and heaviness, by pure vegetable starches, which do a better job of thickening cold sauces. And yet these starches do not wholly remove the problem of starch retrogradation, as it is known, which occurs when moisture seeps out from a sauce as the starch molecules slowly combine, forming a crystal that displaces the water in the starch. A little later chefs perfected the art of making a sabayon, where the thickening action of the egg yolk depends on the formation of microscopic aggregates in the liquid (classically, a sweet wine) in which it is heated. Then came the fashion of nouvelle cuisine, in which sauces were thickened still more gently, particularly ones whose viscosity in traditional preparations is due to the high concentration of gelatin, slowly extracted by long cooking from meat and the bones to which it is attached.

Let's be frank: for quite a long time now, French cuisine has been beating around the bush. Why shouldn't we go directly to the heart of the matter and thicken sauces by using the various compounds that are now available to us in order to obtain new and, above all, more varied consistencies than before? Why shouldn't we harness recent advances in food technology, which already has managed to resolve a number of long-standing problems? In the past, cooks often found that in trying to make solid foods from liquid milk, the milk turned sour. Today, by selecting more suitable gelling agents than gelatin, this form of disaster can be avoided. The same goes for the gelification of wines, which no longer become cloudy once gelatin is replaced by agar-agar, for example. There is no reason that culinary artists should be deprived of simple tools of this sort for creating new dishes. As Brillat-Savarin famously and rightly said, "The discovery of a new dish does more for human happiness than the discovery of a star."

NOT EVERYTHING HAS TO BE SOFT

To say that note-by-note cooking does nothing more than combine compounds is therefore mistaken. One also hears it said that this style of cooking has no consistency, that it's liquid. The same criticism was leveled, no less unjustly, against molecular cooking, which some food critics reproached for its reliance on gels and foams. Understanding why this objection is groundless will help us avoid falling into still more serious errors later.

To accuse molecular cooking of being soft because it uses gels is a sign of great ignorance, because meats, fish, vegetables, and fruits are gels. Gels, by definition, are solid systems that contain a dispersed liquid. Animal tissues are mainly (50 to 90 percent) made up of water, which does not leak out because it is trapped in cells known as fibers. The fibers are aggregated in bundles, which themselves are grouped together in larger bundles, and so on in still larger and larger bundles until they form muscles. Trapped in muscle fibers, the water in meat is released only when, in the course of chewing, our teeth disorganize the tissue—allowing us to perceive juiciness. The same is true of fruits and vegetables, whose tissues are composed of more or less parallelepipedal cells that trap a considerable amount of liquid as well. Meat, fish, vegetables, and fruits, whether they are raw or cooked, are gels; indeed, virtually all traditional cooking is done with gels. This is why I say that some critics of molecular cooking are uninformed.

They are right, however, to have complained that some chefs, in their boundless enthusiasm for whipping siphons and gelling agents, have abused foams and gels, just as in the era of nouvelle cuisine there were those who delighted in undercooking vegetables and serving unreasonably small portions, forgetting (as Brillat-Savarin also remarked) that the appeal of even the most delicious rarity is less the less of it there is. Still, not all molecular cooks are maniacal proponents of these novel devices and ingredients, and I can testify from personal experience that there is quite enough to chew on at the tables of those moderns who know how to manipulate consistencies in interesting ways. Besides, in the unending quarrel between ancients and moderns, the moderns are bound to win because the ancients will always be the first to die. Molecular cooking is in any case not going to go away anytime soon: alginates, carrageenans, and other gelling compounds are now widely available; siphons are sold even in supermarkets today; and as for gels, well, it would not be going too far to say that they are our daily bread—bread, from the physiochemical point of view, is an expanded gel.

Whether what we are making is a gel or something else is less important than exploring different consistencies, hard and soft, tender and tough, moist and dry, flaky and sticky, crunchy and crispy, and so on. The mistaken belief that note-by note cooking must be soft, even liquid, seems to have something to do with the fact that, in the expression "mixture of compounds," the term *mixture* makes some people think of a substance that is not structured. And yet mixing cement and sand makes a very hard mortar; in the kitchen, mixing the white and yolk of an egg makes something solid once it has been cooked. There is no reason why mixing powders and liquids must necessarily produce a liquid. If you mix sugar—sucrose—and water, a liquid syrup forms, but if this syrup is heated (note-by-note cooking, no less than traditional cooking, depends on the mastery of fire), the water evaporates; and once the cooking temperature exceeds 127°C (about 260°F), drizzling the hot syrup onto an unheated marble surface produces a very hard solid. Does anyone think that a dragée made with hardened sugar is either liquid or soft?

It must be admitted, however, that the expression "mixture of compounds" fails to convey the idea that the intelligent cook is obliged to order, organize, and build. The great Marie-Antoine Carême (1784–1833), cook of emperors and emperor of cooks, saw that cooking has much in common with architecture—even if culinary constructions are to be eaten, not merely admired.

*MARIE-ANTOINE CARÊME, AUTHOR OF
LE PÂTISSIER PITTORESQUE (1828).*

With note-by note cooking, the methods developed by Carême have at last been fully extended to the molecular level: the truly modern cook will construct his materials, organize the materials he has constructed, and, finally, assemble and arrange these materials in the form of dishes. We are now going to see how.

THINKING IN PHYSICAL TERMS

There are many sorts of consistency, ranging from nonviscous liquids to perfectly hard solids. But even though this may seem to suggest the existence of a continuum of states, consistency cannot be reduced to a degree of hardness that becomes progressively greater or less, for consistency exists in several dimensions.

Keep in mind, to begin with, that the majority of traditional foods are made of colloidal material in which various phases are dispersed in other phases. The meaning of the terms *colloidal* and *phase* will become clear if we consider the air in a sealed container. Air is a transparent phase (or type) of matter, invisible to the naked eye. Let's therefore use a magnifying glass, then a microscope—if possible, a mesoscope as well, even a nanoscope—to examine our sample at ever greater scales of magnification. Ultimately what we find is a gas that is rather empty, but yet not altogether without molecular motion. A few numbers will help to focus our thinking. A liter of air at ambient pressure and temperature contains about 1.25 grams of matter, mainly dinitrogen, each molecule of which consists of two nitrogen atoms. These molecules are perpetually in motion, a little like billiard balls, and, on average, any two molecules are a little more than twenty molecular diameters apart from each other.

A liquid phase such as water is more condensed than air, which is to say that the molecules in it are closer to each other. Most solid phases are denser still, with a greater number of possible molecular structures, as we will see later on. For a physicist, the chief difference between liquids and solids has to do with their respective flow properties: a solid preserves its shape when it is set on a surface, whereas a gas or a liquid flows, assuming more or less quickly the shape of the container in which it is placed.

PURE SOLIDS

Here, since we are interested mainly in consistency, let's consider solids with regard to their behavior in the mouth. A solid, from this point of view, is a piece of matter that is more or less easily deformed and fragmented. The simplest solids to study are those that are deformed in proportion to the force that is applied to them.

The phrase "in proportion to" may seem clear enough, but to a physicist it is hopelessly vague. From everyday experience, we know that the greater the pressure applied to a solid, the more it will be deformed. A physicist, however, has something more specific in mind: if one applies twice the pressure, the deformation is exactly two times greater; three times the pressure, it is

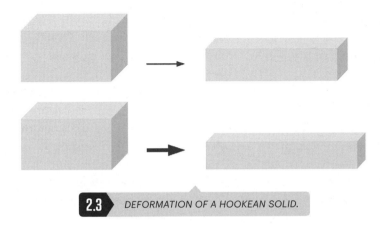

2.3 DEFORMATION OF A HOOKEAN SOLID.

exactly three times greater; and so on. When a physicist speaks of proportional behavior, he really means it. Perfectly proportional behavior is nowhere to be found, of course, at least not in the world we know. But certain solids are nevertheless more perfect than others in this respect. Two basic kinds may be distinguished, Hookean solids and non-Hookean solids.

Hookean solids are ideal solids named after the seventeenth-century English physicist Robert Hooke (1635–1703), who studied the deformation of solid objects, in particular elastic solids that revert to their original shape after being deformed. Not only are Hookean solids deformed in proportion to the force that is applied to them, they also recover their initial form when compression or extension ceases. Their behavior under compression can be described by a single number, the so-called modulus of elasticity (also known as Young's modulus after another English physicist who came after Hooke, Thomas Young). This number expresses the limit of proportionality between stress and strain—that is, between the amount of force applied and the degree of deformation.

Examples of Hookean solids in cooking are at once innumerable and exceedingly rare: innumerable because every (or almost every) solid behaves in Hookean fashion when a small amount of force is applied to it; and exceedingly rare because the force exerted by our teeth in chewing is so great that it actually tears apart the solids that we put in our mouth. Moreover, the food we eat seldom has the form of a solid compressible block. The few Hookean solids we do eat are typically broken up into particles, if not actually crushed into powder, so that we do not break our teeth on them. Sugar is a classic example.

If Hookean solids are rare, then naturally *non-Hookean solids* must be common. Gelatin gels, meats, and vegetable tissues give a partial idea of the range of possible behaviors, from a solid that is more and more easily deformed the more pressure is applied it to one that, by contrast, is deformed less and less easily as the force of compression increases. If you find it difficult to imagine some of these behaviors, think back to the perfect dive and the belly flop: what is true of water is true of solids to a certain extent, especially in the case of gels, many of which consist chiefly of water. (For a physicist, by the way, an aqueous solution—obtained, for example, by dissolving a pinch of salt in a glass of water—is "water." It is not pure water, of course, but the amount of salt is so small that the behavior of the liquid, from the mechanical and rheological point of view, resembles that of water. We may think of it as flavored water. Similarly, melting a fatty substance produces what a cook is accustomed to think of as "oil." I therefore use the term *oil* to refer to any melted fat.)

An entire book might usefully be written about the deformability of solids in the mouth and the way in which they are broken up, for this is what happens when solid matter is placed between the teeth: it is divided, divided again, and then again and again until an equilibrium is finally established between the size of the fragments and the release of saliva, which coats the fragments, facilitating their passage from the mouth to the esophagus. Note that flavor has its origin in this fragmentation. Before a food is swallowed, some of the water-soluble compounds it contains diffuse in the saliva and dissolve (saliva being an aqueous solution of proteins and other compounds). A fruit jelly, for example, which consists mainly of sucrose dissolved in water that has been converted into a gel by the addition of pectins, releases sucrose molecules that dissolve in the saliva. Once these dissolved molecules reach the surface of the taste buds (technically known as papillae), they bind with the molecules of a particular class of proteins known as receptors. The binding of a sucrose molecule to a receptor causes electrical currents—nerve signals—to be propagated to the brain, where they are associated with the sensation of sweetness.

Compounds that are immiscible (that is, not mixable) in water follow a different path. Citral, which has a lovely lemon odor, is one such water-insoluble compound. Instead of dissolving in saliva, its molecules evaporate in the form of gas in the mouth. From there they rise up into the nose, where they bind with olfactory receptors. In either case, the longer one chews, the more intense the sensation. This is why people who eat quickly take less enjoyment from the bioactive compounds present in foods than people who eat more slowly. Their loss—if the food is good!

For the same reason, the friability (or crumbliness) of the solid foods we eat is important. A piece of food that is broken up into a myriad of smaller fragments, each of which releases sapid and odorant compounds, will seem more flavorful than one in which these compounds have a harder time escaping, unless we go on chewing for a longer time. I shall have more to say about all this when the time comes to discuss what may well be called, following Carême, the "architecture of flavor."

Why is a solid more or less solid? So far we have looked at Hookean and non-Hookean solids from the point of view of the physicist, who sees only one aspect of the world. At this juncture it will be helpful to have the perspective of the chemist, for whom the world cannot be reduced to a relatively small number of general laws. The chemist sees it instead as a vast collection of physical systems, each with its own distinctive pattern of atomic and molecular behavior. To oversimplify: the physicist, who believes in the existence of overarching laws that apply *urbi et orbi*, finds to his dismay that he is contradicted from time to time by the particular behavior of different kinds of matter; the chemist, on the other hand, starts from just this, the particular, on which he performs experiments, without, however, always managing to formulate the sort of grand theory that would earn him the fame that is reserved for the most eminent scientists. These two points of view can be reconciled in the form of physical chemistry, which moves from the macroscopic to the microscopic level, starting with a solid object, for example, and proceeding to analyze its molecular and atomic chemistry.

Let's start by looking at a small heap of granulated sugar. It consists of tiny objects that can be clearly seen only with the aid of a magnifying glass. Individually, the grains are transparent: it is the reflection of white light on the faces

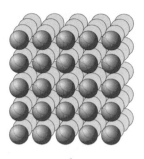

2.4 THE STRUCTURE OF A CRYSTALLINE SOLID.

that makes them appear white; the smaller the grain, the more numerous the reflections. In speaking once again of "faces," I am referring to the fact that the magnifying glass allows us to see that the grains are not rounded, but instead have flat surfaces. An even better way of viewing these facets is to make a large single crystal (or "monocrystal"). Stick a sugar crystal to the end of a thread, dip it in a solution of sweetened water, and let the moisture slowly evaporate. After a few days, you'll wake up to find large a sugar monocrystal with quite visible and regular faces. This monocrystal is formed by the accretion of sucrose molecules along each of the three spatial dimensions, stacked on top of one another like the squares of a series of superimposed checkerboards.

Now dissolve some sucrose in water and heat it in order to eliminate the water. Then, without degrading the sucrose molecules, drizzle the syrup onto a room-temperature marble surface. The abrupt cooling prevents the molecules from moving rapidly enough to become regularly stacked: they remain fixed in the same disordered positions they occupy in the liquid, only now the liquid is a solid. The result is a substance not unlike ordinary glass in a window pane. Because the sugar glass obtained from concentrated syrups crystallizes after a certain time, candy manufacturers use anticrystallants such as glucose to prevent this from happening.

The fact that solids can be either crystalline (exhibiting a regular molecular structure in three dimensions) or amorphous (exhibiting a random molecular structure) does not explain why solids are hard or soft, however. To answer this question we must consider the forces that operate among individual molecules. In a sugar crystal, for example, the sucrose molecules are held together by forces much stronger than the ones that hold triglyceride molecules together in crystals of cooled oil (think of the crystals that appear in bottles of oil stored in unheated rooms during the winter).

Sugar crystals are a type of molecular solid, made of stacked molecules. Salt crystals form another type, the so-called ionic solids. In these crystals, chlorine and sodium atoms are very powerfully bound together, having exchanged an electron (the sodium atom, which normally has eleven such charged particles, gives up one of them to the chlorine atom, which normally has seventeen). The effect of this exchange is to create a stacked molecular structure in which atoms of each kind alternate.

Then there is the class of substances known as polymers, in which so-called covalent bonds assemble the atoms of small molecules into very large, macroscopic molecules. Polymers have a number of interesting and distinctive properties that can be exploited by cooks to create contrasting consistencies. One example is a dish to which I gave the name "Sens dessus dessous" (after Jules Verne's story "The Earth Turned Upside Down"). It is based on a very simple idea, that when we eat we are aware chiefly of the surface of foods. (I applied it to the case of a liquid foam and two solids, but it works

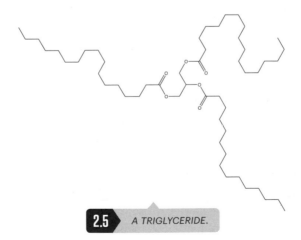

2.5 *A TRIGLYCERIDE.*

with two juxtaposed solids as well.) Begin by making a sort of sandwich consisting of some whipped cream placed between two thin pieces of chocolate. When you bite into it, your teeth come into contact first with the hard chocolate layers. Since our teeth are pressure sensors, we have a sensation of hardness despite the gradual softening of the chocolate by the cream. Now, using the same ingredients, see what happens when you put a piece of chocolate in between two layers of cream. Your initial impression is that the preparation is soft—an impression that is not contradicted when you finally bite into the chocolate. The surface of a food is what captures our attention. Even if cooks limit themselves to solids, they will have no shortage of opportunities for experimentation. After all, there are so many different kinds of hardness!

PURE LIQUIDS

Liquids are no less varied and numerous than solids. Everyone knows that some liquids flow easily (water is the most obvious example), and others less easily (think of honey). What is less well appreciated is that viscosity, a familiar characteristic that physicists have transformed into a quantitative concept, is something more than a parameter of proportionality between gravity (that is, the force that causes a liquid to flow) and the rate of flow. Once again this notion of proportionality—unavoidable in categorizing a great many physical behaviors!

Fill a container with liquid, and on the surface put a thin plate. If you push the plate sideways, the liquid on which it is floating is carried along with it on account of the bonds formed between the molecules of the plate and the molecules of the liquid, as well as among the molecules of the liquid themselves. Obviously if the liquid is nonviscous, like water, friction among the molecules of the liquid is minimal, and only a weak force is needed to move the plate downward. In the case of honey, a stronger force is needed.

Recall once again the comparison between a perfect dive and a belly flop. The rate of speed at which we try to drag the plate is a factor: if the speed is great enough, the water's behavior is no longer that of a liquid, but that of a solid. In other words, while proportionality obtains so long as the plate is

| 2.1 | *VARIATIONS OF DENSITY AND VISCOSITY AS A FUNCTION OF TEMPERATURE* |

TEMPERATURE (°C)	DENSITY	KINEMATIC VISCOSITY 10^{-6} m²/s
5	1.000	1.520
10	1.000	1.308
15	0.999	1.142
20	0.998	1.007
25	0.997	0.897
30	0.995	0.804
35	0.993	0.727
40	0.991	0.661
50	0.990	0.556
65	0.980	0.442

Source: *Hydraulic Models* (ASCE Manual of Engineering Practice No. 25), reproduced in Ranald V. Giles, Jack B. Evett, and Cheng Liu, *Fluid Mechanics and Hydraulics,* 3rd ed. (New York: McGraw-Hill, 1994).

dragged over the surface at low speeds, proportionality is lost past a certain threshold. Still, the behavior of most liquids, when they are made to move at very low speeds, can reasonably be called "perfect" or "Newtonian."

NEWTONIAN LIQUIDS

In the kitchen, as elsewhere, Newtonian liquids (more or less sugary syrups are perhaps the most common instances) are defined as liquids for which deformation is proportional to force. This simplicity is rapidly complicated in cooking, however, because viscosity varies with temperature. In the case of water, for example, table 2.1 shows how density (which makes it possible in principle to create layered cocktails and other liquid preparations, with the denser layers lying beneath the less dense ones) and viscosity (more formally, kinematic viscosity) vary.

Here the temperature is given in degrees Celsius, a unit familiar to most cooks. Density has no unit because it is determined in relation to the density of water at a temperature of 4°C (39.2°F), and the unit of viscosity has no importance in the present context. Only the observed variation matters: when water at room temperature is heated to more than 60°C (140°F), the viscosity is roughly halved. In cooking, the same phenomenon is encountered in many other liquids apart from water, as every chef is well aware.

NON-NEWTONIAN LIQUIDS

Alongside Newtonian liquids there are more complex liquids, such as mayonnaise and (among nonfood products) modern paints. Mayonnaise is fluid in the mouth because it is made of liquid oil droplets dispersed in a slightly acidified aqueous solution. At rest the sauce is firm, however, to the point that a spoon inserted in it stands up, and a dollop of mayonnaise put on a plate seems solid (even if it flows when stirred). The same property is exploited in paint manufacture. Paints today are made to flow easily enough that they can be smoothly rolled on a wall, for example, but not so easily that they drip when rolled on a ceiling.

Mayonnaise and modern paints are examples of what are called rheofluidifying liquids: the prefix *rheo-* means to flow, and *fluidifying* means that the liquid becomes less viscous as it flows; indeed, the more rapidly such liquids are made to flow, the more fluid they become. This is obviously a wonderful property in cooking because in the mouth a quasi-solid suddenly acquires a marvelously smooth quality. Cooks are nevertheless wise not to rely on this effect too often. We also like to eat foods that we can bite into, something with a crunch to it, which is why they should also investigate the properties of rheothickening liquids. Here again *rheo-* refers to the fluid behavior of the liquid, only now it becomes thicker when it is shaken or stirred. In cooking, this effect may be produced by mixing corn starch with water, for example. In sufficiently great concentration (a few tablespoons of water for 100 grams of corn starch), the liquid behaves in a quite remarkable way. If you happen to have a wading pool and enough corn starch, see what happens when you pour some in the pool: walk slowly through the mixture, and it behaves like an ordinary liquid—your feet sink into it; try to run, however, and the pressure exerted by your legs is now much greater, causing the mixture to seem like an impenetrable solid—you have the feeling you're walking on water! Or try shaping a rheothickening liquid into a ball and throwing it up into the air: if you catch it and then immediately throw it back up in the air, it behaves like a solid; so long as it is kept constantly in motion (in a game of catch, for example, or by juggling), its appearance remains that of a ball. But if you throw it to a friend without warning him beforehand what sort of ball it is, he will catch it without throwing it back, and the ball will suddenly turn back into a liquid and run all over his fingers!

Not all liquids are as simple as Newtonian liquids, then. Let's leave it at that for the moment, while taking due note of the fact that they can be used to make striking contrasts. Compare a cup of normal drip-brew coffee and a cup of espresso: the presence of a foam on the surface of the espresso by itself makes an immense difference, even though the foam is made of a gas and a liquid.

COLLOIDS AND DISPERSED SYSTEMS

In the preceding section the temptation to complicate the discussion of liquids was great—and indeed I succumbed to it twice: first by mentioning mayonnaise, in which droplets of oil, a liquid, are dispersed in a liquid phase constituted by egg yolk and vinegar (no mustard in a mayonnaise, by the way, since in that case it becomes a remoulade), and then by imagining a game of catch with a ball made of corn starch mixed with water.

We saw that the behavior of these systems corresponds to that of a liquid or of a solid depending on the proportions of the various phases. If a mayonnaise is liquid when the proportion of oil is small, as you begin to make it, it becomes firm enough to cut with a knife as it approaches the limit at which the emulsion breaks down (because the egg yolk can't absorb any more oil). Let's take a closer look, starting with a pure liquid—that is, a collection of identical molecules in which attractive forces are not strong enough to hold them in place. Then let's dissolve more or less large objects in it, everything from ions (as in the case of salt) to small molecules (sucrose, for example) and then large molecules (not only the soluble component of starch, amylose, but also proteins). These larger molecules are of special interest because they don't quite dissolve; instead they disperse in the form of particles of differing size. As in the case of corn starch, the consistency will completely change depending on the various phases of a system and on the nature and size of the dispersed particles.

Particles? Flour consists of particles—granules. Oil droplets in an emulsion are no different. Now, when such objects are dispersed in a liquid, one obtains a colloid. Historically, the term *colloid* has been used to refer to gels. We shall soon see, however, that gels, emulsions, foams, and aerosols are all closely related. And just as liquids display behaviors whose variety makes the world of sauces a realm of marvels, and just as pure solids exhibit a great diversity of mechanical characteristics, so too colloids inhabit their own world of physical properties. With regard to the fluid mechanics of colloidal substances, in particular, there is a whole range of behaviors between the viscous and the elastic—whence the term *viscoelasticity*.

In exploring this world it will be useful to follow the example of the French philosopher Abraham Moles (1920–1992), who devised a general method of analysis based on what he called "invention matrices," or tables. A table is a marvelous thing, because each cell is an empty space asking to be filled. Up until now we have considered solids, liquids, and gases. The following table (2.2), in which the forward slash / means "dispersed in," summarizes the result of dispersing each of these phases in another phase.

Although it may not be apparent at first sight, the information in this table needs to be restated. Consider the advances in mathematics made possible by the new formalism introduced by Descartes and Leibniz in the seventeenth and eighteenth centuries. Before them, in dealing with equations, it was customary to say, for example, "The square of the unknown multiplied by five and added to the unknown multiplied by three and to the constant two is equal to zero." Today any high school student knows how to express this idea more concisely: $5x^2 + 3x + 2 = 0$. Similarly, the father of modern chemistry, Antoine Laurent de Lavoisier (1743–1794), simplified the language used to describe reactions. Instead of writing, "Ferrous chloride reacts with sodium

2.2 SIMPLE DISPERSED SYSTEMS

A. BASIC CLASSIFICATION INVOLVING THREE PHASES

	GAS	LIQUID	SOLID
Gas	**Gas/Gas:** gas	**Liquid/Gas:** liquid aerosol	**Solid/Gas:** solid aerosol
Liquid	**Gas/Liquid:** foam	**Liquid/Liquid:** either a liquid or an emulsion	**Solid/Liquid:** liquid suspension
Solid	**Gas/Solid:** solid foam	**Liquid/Gas:** gel	**Solid/solid:** solid suspension

Note: The phase in the top row is dispersed in the phase in the left-hand column.

B. A SHORTHAND WAY OF REFERRING TO SUCH SYSTEMS

	G	L	S
G	G/G	**L/G:** O/G or W/G	S/G
L: O or W	**G/L:** G/O or G/W	**L/L:** O/W or W/O	**S/L:** S/O or S/W
S	G/S	**L/S:** O/S or W/S	S/S

hydroxide to form ferrous hydroxide, which precipitates," one now writes, more concisely: $FeCl_3 + 3NaOH\ Fe(OH)_3 + 3Na^+ + 3Cl^-$.

We can do much the same thing. Rather than waste time spelling out the word *gas*, let's simply uppercase the initial letter (G); so too in the case of the terms *liquid* (L) and *solid* (S), as in table 2.2B. Furthermore, the example of mayonnaise reminds us that we should distinguish water and water-based liquids (W) from oil and other melted fats (O). Using these extra letters, we can enrich the informational content in table 2.2A, giving a more precise description where before there was only L.

By a simple manipulation of symbols, then, it becomes clear that in table 2.2A we had omitted to indicate that aerosols can be made from both oil-based and aqueous solutions, that gels can be made from both oil and water, that suspensions can be created in both oil and water, and so on.

Moreover, dispersions of different phases in a single phase can be indicated using the plus sign, as in the expression (G + O)/W, which means that both gas bubbles and oil droplets have been dispersed in an aqueous solution. How? For example, by beating an egg white (an aqueous solution that is roughly 90 percent water and 10 percent proteins) with oil: the whisk divides the oil into droplets while at the same time introducing air bubbles. Thus one obtains a system that might be called either "foamy emulsion" or an "emulsified foam." Neither one of these names is quite appropriate, however, since each assumes a principal system and a secondary system. The formula (G + O)/W does not have this disadvantage: the order of the terms within the parenthesis is determined by an arbitrary rule of alphabetical sequence.

The symbols / and + are called operators. For purposes of geometrical description, it will be helpful also to use the symbol @ to designate inclusion, the symbol σ to designate superposition, and the symbol × to designate imbrication (which is to say the interpenetration of two continuous phases, as in a gelatin gel or a jam). Gels of this kind can be described either as a solid network (or "matrix") extended in an aqueous solution or as water trapped in a solid network. Ought one therefore write S × W or W × S? Here again let's apply the rule of alphabetical sequence and write S × W in referring to such systems.

What value does this theoretical exercise have in cooking? Since we have no difficulty in recognizing solids, liquids, and gases, let us start with these materials—and look to painters for inspiration. How do they grind colors? Using the modern equivalent of a pestle and mortar, they form a paste by dispersing pigments in a siccative oil, so called because it has the property of drying (unlike the oil used in cooking, which remains liquid). In the kitchen, then, begin with sugar and salt since they are solids (even in powdered form, each granule is a solid). If you put the sugar and salt in a mortar and use a pestle to grind them together with oil (a flavored oil, for example), you obtain a liquid suspension or paste. Evidently it is possible to produce either an oil-based paste or a water-based paste (so long as the solid is not water soluble). Alternatively, instead of grinding salt and sugar, one might grind titanium dioxide, a pigment used by pastry chefs to color the surface of white cakes. There is no technical obstacle in this case (although the challenge of

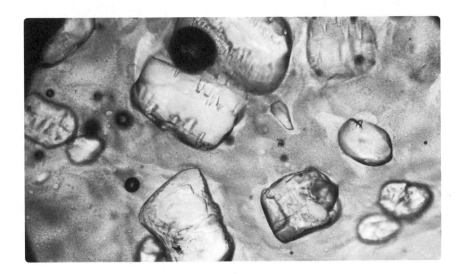

2.6 *A DISPERSED SYSTEM OBSERVED UNDER THE MICROSCOPE.*

reducing titanium dioxide to particles that are nanometric in size—that is, on the order of a billionth of a meter—is apt to give pause to even the most intrepid culinary experimentalist), but there may be a legal obstacle one day. For the moment, under the regulations that govern the food industry in the European Union and Switzerland, titanium dioxide is an approved coloring agent (identified by the code E171). But who can be sure that the health police, in their overzealous determination to turn the world of cooking into a sort of antiseptic retirement home, will not yet succeed in preventing us from using such additives?

Still, we ought not be too pessimistic. Think how much practical knowledge cooks have already accumulated over the centuries! Chocolate makers, for example, have long known that solid particles cease to be perceptible in the mouth if they are too small (strictly speaking, when their diameter is smaller than 15 micrometers, or fifteen thousandths of a millimeter). This, as it turns out, is roughly the size of the particulate matter yielded by conching, a process in which cocoa butter and sugar are slowly ground into a fine paste.

THE PARTICULAR CASE OF EMULSIONS

From the culinary point of view, one of the most important phenomena responsible for giving consistency to liquids, and particularly to water, which accounts for the bulk of vegetable and animal tissues, is emulsification. When it is harnessed by the cook in the proper way, so that the droplets of the dispersed phase are either very finely divided or highly concentrated, emulsification produces consistencies that do not appear to be liquid. We saw earlier that a borderline liquid, which is to say one that flows only beyond a certain threshold, is very close to being a solid.

Whereas aqueous solutions are, well, aqueous solutions, we shall see that oil-based solutions may assume several different forms. Still, let's begin with aqueous solutions in order to familiarize ourselves with some of the compounds that can be used in note-by-note cooking. In traditional cooking, the juices and liquids derived from fruits, vegetables, and meats—wine, broth, stock, and so on—are all aqueous solutions, laden to one degree or another with sapid, odorant, and nutritive compounds. Wine, for example, is an aqueous solution containing about 10 percent ethanol (the alcohol of wine and brandies), mineral salts, tartaric acid, various amino acids, and other organic substances such as phenolic compounds, often wrongly called "tannins" (certain phenolic compounds are in fact used for tanning purposes and so qualify as true tannins, but others aren't and therefore don't deserve the name). What about beef broth and other kinds of bouillon? They consist in the main of water and amino acids, which come from the slow degradation of gelatin, itself obtained by the degradation of collagenic tissue. Bouillon obviously contains many other compounds as well, such as glucose, extracted during the cooking of meat in water, but for the most part it consists of amino acids. All cooks should make a point, by the way, of tasting monosodium glutamate, which is nothing other than the "salt" (as chemists call it) produced by combining glutamate and sodium. Glutamate is an ionized form (obtained, as we saw earlier, through the exchange of an electron) of glutamic acid—an amino acid that is present in meat and has the flavor of bouillon!

Oils exhibit greater diversity, not least because the name "fat" (or "fatty matter") is given to almost any compound that is not soluble in water. Many fats are made of triglycerides. This is true not only of cooking oils, but also of solid fats (animal fats such as lard, milk, cream, butter, and so on). The class of fats nonetheless includes other compounds, such as lecithins, which compose a good part of an egg's yolk, and monoglycerides and diglycerides, which the food industry has long manufactured for use in making emulsifiers.

Lecithins are found in the membranes of human cells, and indeed in the membranes of all animal and vegetable cells. An egg yolk contains many kinds of lecithin, but industrial producers prefer to extract them from soybeans. The commercial use of these products is regulated by law. In Europe, lecithins constitute a class of approved additives (identified by the code E322), and, with the triumph of science over fear represented by molecular cooking, they have now found their way into the kitchen. Still today, however, too few chefs make use of the sodium, potassium, and calcium salts of fatty acids or of monoglycerides and diglycerides or of emulsifiers falling between E470 and E483 on the list of authorized additives. These products have been used in industrial food processing for many years now, as I say, but cooks have scarcely begun to experiment with them. A whole constellation of consistencies is waiting to be explored.

Let me conclude this very brief discussion of a rather immense topic by pointing out that Albert Einstein himself contributed to the investigation of dispersed systems. Einstein reckoned that the viscosity of hard spherical particles dispersed in a liquid is equal to the viscosity of the liquid plus two and a half times the proportion of the dispersed phase in relation to the total volume—a calculation that has been refined since. Do cooks really need to worry about such details? Perhaps not. At a minimum, however, cooks must know that an emulsion cannot absorb more than 95 percent of the dispersed phase: beyond that, sauces break. Two further observations. First, assuming constant proportions of oil and water, vigorously whisking an emulsion makes it easier to smell the odorant compounds dissolved in the oil, whereas in the same emulsion, whisked less vigorously, water-soluble odorant compounds are more easily detected. Second, contrary to what Einstein's calculation seems to suggest, the viscosity of an emulsion completely changes depending on the size of the droplets: a mayonnaise that has been beaten very vigorously is firmer than the same mayonnaise beaten less vigorously.

FOAMS

Although emulsions are indispensable in cooking, they have the disadvantage of enlarging the love handles that naturally grow on our hips with age! The idea of replacing the fatty matter in emulsions with air opens up another vast world for us to explore: foams.

The best known of the foams used in cooking is obtained by vigorously beating an egg white. Whisking disperses the air bubbles, while at the same time breaking them apart into smaller bubbles. When the bubbles are so tightly packed that there is no more room left, they become polyhedric, and the foam firms up. If we add some sugar to the egg white, whisking dissolves the sugar in the aqueous phase that constitutes the walls of the bubbles, and the diameter of the bubbles diminishes as we go on beating the mixture—by much more than it would in the absence of sugar. The technical explanation of this phenomenon is a bit involved, but pastry chefs have long known that it has something to do with the smoother texture that comes from beating sweetened egg whites, similar to the texture that one hopes to achieve in a meringue. Sweetened or not, foams, like emulsions, display a certain threshold behavior; that is, they do not flow easily because the water is trapped between gas pockets by the walls of these pockets. For the moment, to keep things simple, we may speak simply of "capillarity."

Note that compounds other than sugar can be added to the aqueous phase of a foam. Citric acid, for example, contributes a nice lemony flavor; anthocyanins, which give red and black fruits their colors, impart a pleasing hue. The possibilities are immense once you know how to use hydrosoluble compounds. Hydrophobic compounds are a different story. They cannot be dissolved in a water-based foam because, by definition, they are not soluble in water. Why not then disperse droplets of these compounds in the foam's aqueous phase? Nothing is simpler than to beat an egg white until it is stiff and then, still whisking, add to it drops of a flavored oil, an essential oil. This time the foam will have flavor from the molecules dissolved in the aqueous phase, and odor from the compounds present in the oil phase. Another option, of course, is to use hydrophobic compounds in oil-based foams.

Keep in mind, finally, that foams can be cooked. For example, beat a mixture of water and egg proteins (known to pastry chefs as egg-white powder) until it is stiff, then whisk in some beet juice and heat the enriched mixture in a microwave oven for a few seconds, and you will see the foam expand—a sign that the temperature at which water boils (100°C) has been reached. At this temperature, the egg proteins coagulate, and the foam becomes a solid that can be left to cool like a meringue.

PARTICULAR CASES OF GELS

We ought to take a closer look at gels since, as we saw earlier, they are at the heart of traditional cooking: vegetable and animal tissues are essentially dispersions of a liquid in various solids (fruits, vegetables, meat, fish). Using the formalism I have just described, we may refer to these gels by the formula L/S.

We saw earlier, too, that liquids are of two types: water (W) and oil (O). From the formula L/S, then, two others can be derived: O/S and W/S. Now, if we can write down these formulas, we can also put them to use in the kitchen. A gel corresponding to the first of these two formulas is obtained by dispersing some oil (flavored, by all means) in an egg white and then heating the resulting emulsion in a microwave oven as before. After a few seconds it will expand, a sign not only that the water is beginning to evaporate, but also that the temperature has reached 100°C and the egg proteins have coagulated. Take the cooked emulsion out of the oven; this is what I call a "gibbs." Left to dry, it makes an oil gel, O/S, which I call a "graham."

To apply the second formula, let's use sodium alginate. Start with tomato juice and add to it a calcium salt (in this case calcium citrate) recovered from the liquid that coats a clean egg shell after it has been dipped in lemon juice and left to effervesce, which is to say to give off gas bubbles. Next, dissolve some sodium alginate in a bowl of pure water, using an immersion blender to thoroughly dissolve the powder (allowing about 10 grams for 100 grams of water). Now drip the calcium-salted tomato juice into the aqueous alginate solution: beads of liquid form, similar in appearance to salmon eggs, with the liquid interior being sealed by a layer of jelly. Finally, coating these "pearls"

with a gelatinous liquid so that they stick together, we obtain an artificial vegetable tissue corresponding to the formula W/S, to which I have given the
less cumbersome (and, it seems to me, rather prettier) name "conglomèle."

Another group of gels having two interpenetrating phases is designated
by the formula L × S. Here again, two types of liquids can be distinguished,
corresponding to the formulas O × S and S × W. The second has long been
familiar in the form of gelatin gels and jams. The first, by contrast, is new in
the history of cooking.

Still more possibilities will come into view at this point if we introduce
the concept of dimension. The basic idea is quite simple. Remember the sort
of cube we all played with as children? If you put the cube in front of you
and then change your own position in relation to it—moving from the left
near corner by a certain distance to the right, then by another distance to
the rear, and, finally, by a third distance toward the top—you can locate any
point inside the cube by reference to three numbers. Three numbers, three
dimensions. To locate all the points on a sheet of paper, however, only two
numbers are needed because the sheet has two dimensions—assuming that
it is infinitely thin (a crucial assumption—otherwise we would find ourselves
back in the previous case, with one dimension, thickness, being smaller than
the two others). An actual sheet of paper does have a thickness, of course.
But isn't it "almost" zero? Everything depends on what are known as orders
of magnitude. Suppose we have a sheet that is square, with sides 20 centimeters in length, and whose thickness is a millimeter—200 times less than 20
centimeters. A number a is said to be smaller than another number A by an
order of magnitude if the ratio A/a is equal to or greater than 10. For our sheet
the ratio is 200, so the thickness is smaller than the length of the side by two
orders of magnitude.

The notion of an order of magnitude makes it possible to understand what
is meant by the dimensionality of an object. In the case of a sheet of paper,
because the dimension corresponding to thickness is smaller than the other
two dimensions by more than an order of magnitude, the sheet is conventionally said to be of dimension two. Similarly, a string—in the kitchen, a spaghetti
noodle, for example—is of dimension one because two of its dimensions are
smaller than the third dimension by at least an order of magnitude. Thus, too,

2.7 *A LIEBIG, BY PIERRE GAGNAIRE.*

a grain of rice may be said to be of dimension zero if the plate on which it sits is taken as the basis for measurement.

Let us now go back to the operator /, which signifies random dispersion. We could, of course, limit our attention to a formula such as W/S, in which a liquid is dispersed in a solid. But we can speak more precisely by considering the dimensions of gases, liquids, and solids: instead of indicating simply the nature of the substance dispersed, we can note its dimensionality (D) as well. One might write D_0/D_3, for example, or D_1/D_3 or D_2/D_3. What do these formulas mean? The first, D_0/D_3, refers to small, zero-dimensional objects dispersed in a three-dimensional mass. (In the case of a gel, the dispersed zero-dimensional objects are liquids.) The second formula, D_1/D_3, corresponds to strings dispersed in a volume, and the third, D_2/D_3, to sheets dispersed in a volume.

How does one go about producing such objects in the kitchen? It will be obvious that the first formula, D_0/D_3, describes vegetable tissues—natural and artificial alike. Earlier, I described making conglomèles in the form of

2.8 *A FIBRÉ, BY PIERRE GAGNAIRE.*

liquid-filled sacs (or "pearls"), one by one, but there are many ways of doing this more quickly. For example, you could stack small pieces of frozen liquid (lobster bisque, say) in a box and then fill it with a gelatinous liquid (grapefruit-flavored chicken stock, perhaps, in which some gelatin has been dissolved). The frozen liquid will melt as the stock gels, and you will be left with a mass of pearls dispersed in the gel.

How about dispersing a fatty substance, rather than an aqueous solution, in a solid matrix? It would then be possible to obtain the preparations to which I have given the names "liebig" and "gibbs," respectively, by emulsifying a fatty liquid in a solution enriched with gelatin or by emulsifying it in an egg white (in which case the emulsion is heated).

As you can see, giving material form to a formula is quite straightforward. The second of the three formulas mentioned, D_1/D_3, is also simply applied. For example, if the one-dimensional structures are liquid, then the formula describes animal muscle tissues (meat, fish). These physical systems can be recreated with the aid once again of alginate. Take a tube of calcium-enriched aqueous solution and let it gel in a tray containing water in which sodium alginate has been dissolved. This will give you long strings filled with water. If you then arrange the strings lengthwise and form them into a cohesive bundle, you will obtain a piece of artificial meat, to which I have given the name "fibré." Here, too, other methods are possible. For example, the first time I made a fibré, I put long hollow noodles in a glass and then filled it to the top with a hot aqueous solution in which some gelatin had been dissolved.

Alternatively, one might cut a lattice of cylindrical shafts in a solid substance (using, say, a board of wood with long protruding nails that, when pressed into the solid, make a matrix of holes in one go), then fill the shafts with a liquid and plug them.

THE MECHANISMS OF COAGULATION

Up to this point, I have considered the structure of gels, but not their molecular aspect. We need first to distinguish two principal types of gel, physical and chemical. To make the first type, dissolve some gelatin in warm water; as the solution cools, the gelatin forms a lattice that traps the water. The water no longer flows, and we are left with a particular kind of solid, a gel. Once heated it melts and again forms a liquid that, on cooling, will reassume the form of a gel. The temperature at which this transition takes place is rather low, about 36°C (97°F). In other words, not much energy is needed to dissociate the molecules of a gel of this type because the bonds between gelatin molecules are weak. The gel is thermoreversible: it is a physical gel.

Another example, this time a chemical gel. An egg white is composed of water in which proteins are dissolved (accounting for about 10 percent of the total mass). The liquid solidifies when it is heated, but the more it is heated, the harder the gel becomes. Moreover, the gel is thermoirreversible; that is, it cannot be converted back to its original state through heating because the proteins degrade before separating. (Some years ago, I showed how to "uncook" an egg by separating the protein molecules, but this process requires a much more energy-intensive method than simply heating to 36°C.) Eggs, moreover, are not the only culinary ingredients containing proteins that coagulate to form a chemical gel. Meats, like fish, contain great quantities of protein. This is why we can make terrines: ground meat (which can be stretched by adding water) forms a gel when it is heated.

The use of proteins to structure liquids will become clearer if we examine a cousin of muscle tissue, the artificial fish-based food product known as surimi. Manufacturing surimi involves two basic operations. The first is carried out at sea, on board a factory longliner (the industry name for a commercial trawler equipped with a processing plant), where fish are caught, headed, and gutted. The filets are then thrown together, washed several times in fresh water, and pressed. Eliminating the soluble proteins, blood, fat, and connective tissues yields "surimi base"—a white tasteless paste, rich in proteins and poor in lipids. Compounds are then added to improve the proteins' resistance to cold (polyphosphates, sugar, sorbitol), and the paste is cut up into 10-kilogram slabs that are then frozen to –30°C (–22°F). When the factory longliner returns to land, the second phase of processing takes place. The frozen surimi base is thawed on arrival at the production facility and inspected for whiteness, gel cohesion, microbiological content, and so on. Various substances are then added to it: starch (potato or wheat), egg white, oil, salt, sorbitol, calcium sulfate, as well as natural or artificial fragrances (crab, shrimp, lobster) and coloring agents (paprika is often used to give the final product an orange tint on the outside). The paste is then kneaded and rolled out to form a thin layer and steamed. Finally, it is shaped (usually rolled and scored), vacuum packed, and pasteurized.

Is surimi any good? Without quibbling over semantics ("good" is, in any case, a very personal judgment), it must be admitted that surimi is highly variable in quality. Selecting the right ingredients is important: if a low-quality fragrance is used, the odor will be uninviting; if an excessive amount of starch is used, the texture will be doughy. Everything depends on how much intelligence and labor are put into making it—and also how much money ("The cost is forgotten, but the quality remains," as the gangsters say in the old French film *Les Tontons flingueurs* [Crooks in Clover, 1963]). From the chemist's point of view, the wonderful thing about surimi is that it shows how easily proteins can be strung together to produce quite extraordinary consistencies. What's more, it is just one possibility among a vast number—further proof that note-by-note cooking is not limited to making jellies, jams, and gelatins!

ADDITIVES

More than once in our exploration of consistencies we have found ourselves following in the footsteps of the food industry, which in turn was following in the footsteps of cooks. Cooks learned long ago how to make emulsions using eggs. Scientists subsequently investigated the properties of eggs. From scientists the food industry learned that proteins and lecithins are good emulsifiers and then found a way to isolate lecithins from inexpensive vegetable matter such as soybeans. Fittingly, then, molecular cooks have closed the circle by substituting artificial emulsifiers when, for one reason or another, the flavor of an egg yolk is unwanted. So, too, scientists succeeded in explaining why cooking down veal shanks makes it possible to make aspics and desserts such as Bavarian cream: the collagenous tissue is dissociated when it is heated in water, releasing gelatin, a natural gelling agent. Today, commercial food manufacturers make the like of these products, originally obtained through a long and costly process of cooking, only now much more efficiently and cheaply by using gelling agents extracted from algae.

To protect citizens (you will forgive my use of a political term here—it's simply that I refuse to call people "consumers"), governments regulate the use of novel products of this sort. In France, regulation is based on a 1905 law stipulating that food products must be "wholesome, genuine, and saleable." It is forbidden, for example, to sell damaged fruits and vegetables or to sell a product under a false or misleading name. All this is as it should be. But it is a sad state of affairs, in my view, when we are led to fear an entire class of compounds that impart not only color and flavor to foods, but also consistency. The desire to give different consistencies to foods is by no means new. Cooks have long made use of egg, meat, and fish proteins for just this purpose, as we have seen. One might even regard traditional cooking as a perpetual oscillation between stronger and weaker forms of consistency. To make a terrine, for example, meat is ground up until it forms a soft paste and then cooked until it forms a hard paste. Carrots, which in their natural state are hard, are softened by cooking, but then become firm again when they are used to make a carrot flan. Are these pointless operations? Not at all—they are ways of creating flavor!

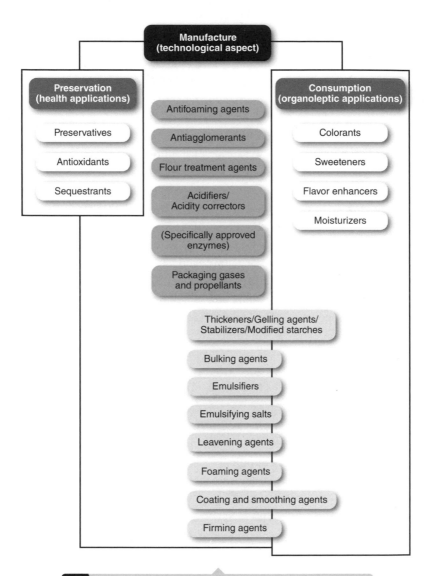

**Manufacture
(technological aspect)**

**Preservation
(health applications)**

Preservatives

Antioxidants

Sequestrants

**Consumption
(organoleptic applications)**

Colorants

Sweeteners

Flavor enhancers

Moisturizers

Antifoaming agents

Antiagglomerants

Flour treatment agents

Acidifiers/
Acidity correctors

(Specifically approved
enzymes)

Packaging gases
and propellants

Thickeners/Gelling agents/
Stabilizers/Modified starches

Bulking agents

Emulsifiers

Emulsifying salts

Leavening agents

Foaming agents

Coating and smoothing agents

Firming agents

2.9 MAIN CATEGORIES OF FOOD ADDITIVES APPROVED FOR
USE IN THE EUROPEAN UNION AND SWITZERLAND.

Gelification and thickening are therefore matters of some importance. Over many decades the food industry created a range of products that today are known as texturing agents. Again, there is nothing new about the basic idea. Cooks have long been accustomed to thicken sauces with flour. They have long known that cooking meat and fish (with or without their bones) yields bouillons that gel, which is to say liquids that can be converted into solids. People who live near the sea have found many uses for algae, wrack, and lichens, whose polysaccharides are the source of various modern texturing agents. Alginates were first used by the cosmetics industry and later introduced in food processing, along with carrageenans (extracted from red algae), xanthan gum (produced by the bacterial species *Xanthomonas campestris*), and so on. Many such agents figure today on the list of officially approved additives, which are divided into a number of categories.

The left-hand and right-hand columns of figure 2.9 need not detain us here. The middle column is the one we need to examine, in particular the lower part, which catalogs the various types of compounds that are useful in cooking. Emulsifiers make it possible to make emulsions, foams, suspensions; emulsifying salts are useful for modifying the consistency of milk-based gels; leavening agents, by releasing gas, produce foams; foaming agents induce expansion; firming agents make food products less soft, thickening agents make liquids more viscous. Let's take a closer at these substances in the order of their E number, the numerical code assigned to food additives approved for use within the European Union and Switzerland.

E400 alginic acid (emulsifier/thickener and gelling agent)
E401 sodium alginate (emulsifier/thickener)
E402 potassium alginate (emulsifier/thickener)
E403 ammonium alginate (emulsifier/thickener)
E404 caloium alginate (emulsifier/thickener)

All of these substances are extracted from brown algae of the class Phaeophyceae (a group that contains the Fucaceae, Laminariaceae, Alliaceae, and Lessoniaceae families). Molecularly, alginates are linear polymers of alpha-L-guluronate acid and beta-D-mannuronate acid.

The last part is a mouthful, I know. Let's take it one step at a time, beginning at the beginning—with glucose. Glucose, as we saw earlier, is a compound. In its pure form, at room temperature, it is a crystallized or amorphous powder. In white light, glucose is white: the light is reflected by its solid granules, which individually are transparent. Glucose has a very mild flavor, not really sweet. And it's literally in our blood, which carries it as fuel for cells throughout the body. From the chemical point of view, glucose comes under the head of sugars (and, more precisely, of monosaccharides), first systematically studied by the brilliant German chemist Emil Fischer (1852–1919) in the late nineteenth and early twentieth centuries.

Like glucose, fructose is a sugar, but it is much sweeter. Both glucose and fructose are found in vegetable tissues, particularly in the elaborated sap that descends from the leaves of plants, where saccharides and, in smaller amounts, amino acids are synthesized. Unsurprisingly, these sugars are also found in the underground organs of certain vegetables (the roots of the carrot and the tubers of the potato plant, for example), since these organs serve primarily to store energy. Glucose and fructose are classified as "simple" sugars, or monosaccharides, because they can be linked together to form larger molecules. Sucrose, for example, is formed by joining glucose and fructose. In onions, somewhat larger sugars called oligosaccharides are formed by several fructose molecules linked to a sucrose molecule.

Glucose and fructose are not the only basic molecular components, or building blocks. Two other elementary sugars, guluronic acid and mannuronic acid, can also form chains. There are many other such compounds as well. When a molecule is made up of fewer than a hundred or so repeating units (known as monomers), it is called an oligomer (from the Greek *oligo-*, meaning "few"). By contrast, when the number of monomers is higher than a hundred, the molecule is known as a polymer (from the Greek *poly-*, "many"). Note, too, that elementary sugars can be joined together at their ends to form a ring, in which case the polymer is said to be "linear," or arranged in the shape of a tree, in which case one speaks of a "branched" polymer.

With regard to alginates, there are three further things to keep in mind: first, that in an acidic environment they precipitate in the form of alginic acid, the initial item on the preceding list (E400); second, that the molecular constitution of a particular alginate in a given marine substance determines its exact properties as well as the appropriate manufacturing technique; and third, that alginate molecules can be tightly bound together with the aid of calcium ions, as we saw earlier with liquid-filled pearls.

E406   agar-agar (emulsifier/thickener)

Agar-agar is not a very new substance, but just the same it appears in the category of food additives because it is indeed "a substance added in small quantities to foods, in the course of their preparation, for a precise, technical purpose," as the official definition stipulates. Agar-agar (often abbreviated as "agar") is extracted from red algae of the class Rhodophyceae. From the molecular point of view, it is a linear polymer whose basic pattern (the elementary ring of the chain) is composed of two galactose residues, more or less modified by various atoms. Agarose, the principal constituent of agar-agar, is composed of beta-D-galactose and a 3.6 anhydro-L-galactose. The length of each molecule depends on its actual source and the particular extraction procedure employed. Extraction is performed by solubilization in water at a temperature of 100°C for several hours in the presence of a base. Separation is done by freezing.

Sorry, I forgot to introduce galactose—a monosaccharide found particularly in milk sugar, lactose, which is a disaccharide like sucrose. Whereas sucrose is made of glucose linked with fructose, lactose is made of glucose linked with galactose.

E407   carrageenan (emulsifier/thickener)

Like agar-agar, carrageenans are a naturally occurring polysaccharide isolated from red algae of the class Rhodophyceae, but this time the basic pattern is a sulfated beta D galactose linked with an alpha D galactose, sulfated or not, sometimes in 3.6 anhydro form.

If anyone is wondering, by the way, whether the mention of a "sulfated" component here should arouse fears of copper sulfate poisoning, the answer is emphatically no. Human cartilage also contains sulfate groups, composed of one sulfur atom linked with four oxygen atoms. Moreover, it is mainly the copper in copper sulfate that is toxic, as you can readily see by means of a simple experiment. Let some water stand in two bottles, one of which contains copper turnings. After a few weeks or months, you will see green algae in the bottle without copper, but none in the other bottle because the copper turnings poisoned the algae before they could become established.

To come back to carrageenans. I use the plural since conventionally one distinguishes between iota-, kappa-, and lambda-carrageenans according to the degree of sulfation and the position of the sulfated carbon atoms. These compounds do not share the same properties, nor are they easily substituted for one another. For example, whereas kappa-carrageenan forms rigid, brittle, slightly opaque gels, iota-carrageenan forms elastic, clear gels, and lambda-carrageenan does not form gels at all, though it does promote viscosity. Preparation of all these products is the same as in the case of agar-agar, and recovery is done by drying the residue of a solution that has been filtered by alcohol precipitation (microballs, formed with the addition of potassium chloride [KCl], can be used instead of alcohol).

E410   carob gum (emulsifier/thickener)

Carob gum (also known as locust bean gum) is obtained from the seeds of the carob tree (*Ceratonia siliqua*), native to the Mediterranean littoral (Spain, Portugal, Greece, Morocco). It is a galactomannan—a linear polymer composed of alpha-D-mannose units, with lateral branching of a single alpha-D-galactose unit. Carob gum comprises on average one galactose unit for four mannose units. (Mannose, as everyone will have guessed by now, is another simple sugar, a monosaccharide.)

Carob seeds are reduced by a series of mechanical procedures (separation of the endosperm, removal of the seed hull, milling) that yield a flour consisting very largely (95 percent) of galactomannans.

E412   guar gum (emulsifier/thickener)

Guar gum comes from the seeds of the guar tree (*Cyamopsis tetragonolobus*). It is produced mainly in India and Pakistan, and to a smaller extent in Texas. Made up of galactomannans, like carob, guar gum is characterized by the repetition of one galactose unit for every two mannose units.

E413 tragacanth gum (emulsifier/thickener)

Also known as adragante gum, tragacanth gum is obtained from several species of *Astragalus*, shrubs of the Leguminosae family. It is a mixture of two polymers: tragacanthic acid, which is a polymer of D-galacturonic acid that also contains D-galactose, D-xylose, and small quantities of L-rhamnose; and a neutral polymer, tragacanthin, which is a highly branched arabinogalactan, the principal chains being constituted of D-galactose residues and branched L-arabinose residues.

E414 acacia gum (thickener/stabilizer)

Better known perhaps as gum arabic, acacia gum is produced in many regions of Africa, generally from several species of the acacia tree, *Acacia senegal* (or *Acacia verek*), *Acacia seyal*, and *Acacia laeta*. It is constituted by densely branched polymers, with a principal chain consisting of linked beta-D-galactose residues, with substituted beta-D-galactose units having arabinose and D-galacturonic acid endings. Already in the nineteenth century, cookbooks mentioned the use of gum arabic as a stabilizer for whipped creams—nothing very new here on the industrial side, then.

E415 xanthan gum (thickener)

An extracellular polyoside, xanthan gum is produced by aerobic fermentation of the microorganism *Xanthomonas campestris* on a glucidic substrate (typically molasses) under specific conditions: pH between 6 and 7.5 (pH is a measure of acidity, ranging from 0, very acidic, to 14, very alkaline) at temperatures between 18° and 31°C (64.4° and 87.8°F). The gum is recovered at the end of the fermentation process by alcohol precipitation, then dried and ground.

From the molecular point of view, xanthan gum is a polymer composed of a linear chain of beta-D-glucose with lateral branching at every other glucose unit. These side chains contain an alpha-D-mannose, a glucuronic acid, and a terminal beta-D-mannose.

E417   tara gum (stabilizer)

Tara gum is obtained from the seeds of the tara tree (*Caesalpinia spinosa*), found mainly in the South American sierra (especially in Peru) and harvested chiefly for the tannins extracted from its pods. The gum is a by-product of this operation.

The polymer in this case is formed by a linear chain of beta-D-mannose, with only single-unit branching of alpha-D-galactose residues.

E418   gellan gum (thickener/stabilizer)

An extracellular polyoside produced by aerobic fermentation of the microorganism *Pseudomonas elodea*, gellan gum is obtained by a method similar to the one employed in the production of xanthan gum. Commercial processing uses two forms of the product: a native, acetylated form and a deacetylated form. The native form is a linear molecule whose basic unit is a chain of glucose residues, glucuronic acid, glucose, and rhamnose. In half of the units, an acetyl group (one carbon atom linked to three hydrogen atoms and to another carbon atom that is itself linked to two hydrogen atoms) is linked to one of the glucose residues. Nevertheless the gelling properties of the esterified form remain limited, and it is the deacetylated form that is more commonly used.

E440   pectins (emulsifier/thickener)

Pectins are complex mixtures extracted from the cell wall of plants. It is these substances that cause jams and jellies to set: cooking fruits and sugars extracts pectins and dissolves them in the acidic liquid released by the fruits as they are broken down; then, on cooling, the extracted pectins join together to form a gel that traps the liquid from the fruit along with bits of its flesh. The food industry found it profitable to look for ways of extracting various kinds of pectin individually, rather than let what was left over from the pressing of fruits (citrus fruits and apples in particular) and the processing of sugar beets go to waste. Pectins are first extracted by solubilization in a heated acidic medium and then, after purification and concentration, precipitated with alcohol, dried, and finally ground.

The major molecular constituent of pectins is galacturonic acid, with a few inserted rhamnose residues. Side chains are attached to a backbone, forming "hairy" regions that alternate with "smooth" regions. The smooth (or homogalacturonic) regions are made up exclusively of linked galacturonic acid units and account for 80 to 90 percent of the molecular structure of pectins.

Varieties of pectin are distinguished because they can be partially esterified to one degree or another, which is to say linked in different proportions with methanol, an alcohol related to ethanol (the alcohol found in brandies, wine, and beer). The relevant parameter in classifying the gelling properties of pectins is the degree of esterification (DE)—that is, the number of carboxylic acid groups in a pectic chain esterified by methanol per hundred galacturonic units. In this way high methyl-ester (HM) pectins, having a DE greater than 50 percent, are distinguished from low methyl-ester (LM) pectins, whose DE falls between 25 percent and 45 percent. Depending on whether more or less sugar is combined with more or less calcium, one can obtain gels of different degrees of firmness. A word to the wise cook: choose your pectins well!

E460 microcrystalline cellulose, powdered cellulose (emulsifier/thickener)

E461 methyl cellulose (emulsifier/thickener)

E462 ethyl cellulose (emulsifier/thickener)

E463 hydroxypropyl cellulose (emulsifier/thickener)

E464 hydroxypropyl methylcellulose (emulsifier/thickener)

E465 ethyl methyl cellulose (emulsifier/thickener)

E466 carboxymethyl cellulose (emulsifier/thickener)

Cellulose is the most abundant organic polymer on earth. Its chemical inertia and insolubility in water make it indispensable in textile manufacturing: If it were to dissolve in hot water, our shirts and blouses would vanish when we wash them. (The cotton used in making clothes is virtually pure cellulose.) Where does it come from? From vegetable matter: vegetable tissues are made of cells cemented to one another by a wall that is made up of cellulose molecules—great pillars, in effect, held in place side by side by a web of pectin molecules.

Cellulose itself holds little interest in connection with thickening or gelling. But some of its derivatives, generally obtained by simple modifications that make cellulose molecules hydrosoluble, have surprising properties. Although most cellulose derivatives are thickeners, there are exceptions: methyl cellulose, for example, gels on being heated and reliquifies on being cooled—the opposite of gelatin and pectins!

E1200 polydextrose (stabilizer/thickener)

Let us now turn our attention to another substance that is naturally very abundant on earth—starch, found in wheat, rice, potatoes, corn, and so on. The name "starch" is given to grains that are composed chiefly of two types of compound: amyloses and amylopectins (the latter have nothing to do with pectins). Amyloses are linear polymers, and amylopectins are branched polymers; in each case the basic repeating unit is glucose.

Why are these polymers present in the vegetable kingdom? A teleological reply to such questions, which assumes a purpose in the evolution of living things, is generally to be avoided (since otherwise one ends up saying that our nose and ears, for example, were created so that we could wear glasses). Nevertheless it is true that glucose, which is synthesized in the leaves of plants, is a water-soluble compound and so liable to being washed out by rain, whereas starch, a water-insoluble substance produced in the course of seed germination, very usefully remains embedded in roots, seeds, and tubercles as an internal reserve of energy.

Cooks have long been adept at thickening soups and sauces with flour, kneaded butter, roux, starchy foods such as potatoes and rice, and so on. The problem with using plant (or native) starches is that they give sauces, especially, a sticky consistency. Sauces containing amylose, such as ones that are thickened with corn starch or wheat starch, likewise form opaque and rigid gels when cooled; and if they are kept warm for a certain time (in a bain-marie during restaurant service, for example), they expel water—a phenomenon known as syneresis, in which moisture beads up on the surface of a sauce because the amylose molecules released during cooking reassociate and form a gel.

In order to avoid these inconveniences, which cooks have found various ways of minimizing over the centuries, the food industry devised a method, known as reticulation, for modifying starches. Reticulated starches, in the form of either distarch adipates or distarch phosphates, are characterized by the presence of cross-linked polymer chains that strengthen the cohesiveness of starch grains, prevent syneresis, and make gels less viscous.

Reviewing the list of thickening and gelling additives, one item at a time, may seem a bit tedious, but only if you do not stop to consider their culinary uses! The compounds I have mentioned so far yield very different gels and are used for a variety of purposes. Their properties are summarized in table 2.4.

Surely the information contained in this table justifies molecular cooking's introduction of such products. Why should cooks be condemned to use only native pectins and the gelatin extracted from veal shanks when so many new alternatives are available? It is as though musicians had limited themselves to the triangle, without touching the piano that was offered to them; as if painters resolved to draw only with a charcoal pencil, even though a whole range of colors had been placed at their disposal!

In recent years, fearmongers—for that, frankly, is what they are—have shamelessly sought to take advantage of a certain mood of ignorant naturalism in order to make people believe that additives are dangerous. But isn't there more danger in a barbequed piece of meat than in xanthan gum? The answer, unquestionably, is yes. And yet, as I know myself from personal experience, arguing on the basis of examples even as telling as this one is useless. The only way of convincing people of the truth is to explain the facts as carefully and as impartially as possible, without seeking to convince. There is a paradox in this, I admit. But there you are. Anyone who wishes to go on cooking as people did in the Middle Ages should by all means feel free to do so. Other, less timid souls will gladly adopt the compounds I have just described. They will throw themselves into note-by-note cooking, seeing in it, as I do, the source of a new kind of connoisseurship. Over time, I predict, the success of note-by-note cooking will erase the altogether arbitrary distinction that is made today between meat, fish, fruits, and vegetables, on the one hand, and compounds (not only additives, but all such examples of chemical

SELECTED GELLING COMPOUNDS

A. CLASSIFIED ACCORDING TO SOLUBILITY, EFFECT OF HEATING, AND GELLING REQUIREMENTS

COMPOUND	SOLUBILITY	EFFECT OF HEATING	GELLING REQUIREMENTS
Agar-agar	Heated	Withstands autoclaving	—
Kappa-carrageenan	Heated	Does not melt at room temperature	Potassium needed
Kappa-carrageenan and carob	Heated	—	Potassium needed
Iota-carrageenan	Heated	—	Sodium, potassium, or calcium needed
Sodium alginate	Heated	Not thermoreversible	Calcium needed
HM pectins	Cooled	—	Sugar and a pH lower than 3 needed
LM pectins	Cooled	—	Calcium needed
Gum arabic	Cooled	—	—
Modified starches	Heated	Cause syneresis	—

B. CLASSIFIED ACCORDING TO TEXTURE, APPEARANCE, AND APPLICATION

COMPOUND	TEXTURE OF GELS	APPEARANCE	APPLICATION
Agar-agar	Firm, brittle	Clear	Confectionery
Kappa-carrageenan	Brittle	Clear	Desserts, custards
Kappa-carrageenan and carob	Elastic	Opaque	Desserts
Iota-carrageenan	Flexible, elastic	Clear	Frozen desserts
Sodium alginate	Brittle	Clear	Desserts and jams
HM pectins	Spreadable	Clear	Desserts
LM pectins	Brittle	Clear	Milk-based desserts
Gum arabic	Soft	Opaque	—
Modified starches	Rigid to supple	Opaque	Puddings, desserts

ingenuity) on the other. And with it, I pray, will come the end of the obligation that presently falls upon the regulatory agencies of the state to introduce
legal niceties that have no rightful place in food and cooking. May the very
notion of additives disappear as soon as possible, in fact—because these compounds, far from being mere technical accessories, form the basis of the very
foods we eat!

CONTRASTING CONSISTENCIES

In his first note-by-note dish, Pierre Gagnaire created what I had the idea
of calling "péligot disks," after the nineteenth-century French chemist Eugène-Melchior Péligot (1811–1890), a sort of glucose caramel that he stacked
in thin round slices, one on top another. The oral sensation produced by three
superimposed disks is wholly unlike the sensation produced by a single disk:
one perceives not so much crunchiness as crispiness, the result of a cascade
of tiny crackles. By itself, this example shows that the basic operations of dispersion, inclusion, superposition, and interpenetration must be reconceived
if we are to obtain sensory effects in the mouth that will improve upon the
flat, vapid, dull quality of a simple gel.

We already have had an opportunity to appreciate the usefulness of em-
ploying various symbols, known as operators, in analyzing the relations
among objects of different dimensions. By defining a restricted set of symbols
that, together with the rules governing their use, constitutes a formalism for
describing dispersed systems, it becomes possible to imagine a vast field of
culinary innovation in terms of consistency alone.

Let's begin by asking how many options are available to us if we wish to
create a dish having only two components. The range of possibilities contemplated by the Dispersed Systems Formalism I proposed several years ago is
given in tables 2.4 through 2.7. Each table considers two objects of a particular dimension and describes the structural consequences of subjecting them
to one of the four basic operations just mentioned—dispersion, inclusion, superposition, and interpenetration. (Obviously, there is no particular reason
why a dish should be limited to two objects, and so one would have to go on
to draw up tables for three objects, four objects, and so on.) Once again there

DISPERSION
(SYMBOL I) **D_0**   **D_1**

D_0

D_0/D_0: First, we must establish a standard of magnitude (or size)—let's say a plate that is 20 centimeters in diameter. In that case, an object is of dimension zero if its dimensions measure less than 2 centimeters. Thus, for example, a cherry tomato is zero dimensional. If smaller objects are dispersed in an object that is less than 2 centimeters in diameter (it need not be spherical), we obtain a structure described by the formula D_0/D_0. In traditional cooking, for example, one thinks of a *Royale Saint-Germain de petits pois frais*—a small cube formed by coating baby peas with an egg batter and then cooking them in a mold. The corresponding dish in molecular cooking is a cubical mass of alginate pearls.

D_1/D_0: It is very difficult to disperse many fibers in a volume less than 2 centimeters in diameter—unless the fibers are short and fine. Asian noodles might be a suitable candidate.

D_1

D_0/D_1: Here objects less than 2 centimeters in diameter are dispersed in linear (though not necessarily straight-line) objects having a diameter less than 2 centimeters. For example, we might arrange walnuts in a row on a brick pastry sheet and then roll it up. We might also cut a gel into small dice and then string them together in a necklace. We now have a marvelous way of making a new texture because it will give us the impression of tasting a succession of flavored beads—an altogether novel sensation, as far as I know.

D_1/D_1: Here one-dimensional objects are dispersed in an object of the same dimension. Observe that every bundle of fibers corresponds to this formula (in this case the diameter of the bundle must be less than 2 centimeters). Nevertheless, there is no law that says that individual fibers must be arranged in a line. We can braid them or disperse them at random in the largest fiber. Have you ever had the pleasure of eating braided spaghetti, for example? A remarkable sensation, I assure you.

D₂/D₀: Think, for example, of crumpling sheets of some edible product into a small ball or other shape.

D₃/D₀: An impossibility—for an object 20 centimeters in diameter cannot be contained in an object only 2 centimeters in diameter. No matter, there are so many possibilities that one more or less hardly makes any difference.

D₂/D₁: Speaking of sheets (or leaves), they can be dispersed in a fiber or in a rope. One obvious method would be to roll up a stack of sheets into a cylinder, like a rolled-up newspaper. Why not also put sheets inside this cylinder that have been crumpled into balls?

D₃/D₁: Here, too, an impossibility.

(CONTINUED)

DISPERSION
(SYMBOL l) D_0 D_1

D_2

$D_0 l D_2$: Here small objects of zero dimension are dispersed in an object of two dimensions. How? "Two dimensional" means that one of the dimensions, thickness, is less than 2 centimeters. One might extend the method of the *Royale Saint-Germain* mentioned earlier in $D_0 l D_0$ by dripping egg batter on a layer of baby peas and then heating the result so that it coheres, forming a layer of pearls—a gourmet's delight!

$D_1 l D_2$: Strings dispersed in a layer: one thinks at once of the art of weaving. It would be fun to weave spaghetti into a tissue, but it would be a great deal of work without a machine. The technique used in making artisanal paper seems better suited to our purpose: cook the spaghetti, then spread them out on a plate in a thin layer and bind them together with a jellied coating. Structures much smaller than spaghetti would obviously be a still better choice. For example, if we were to centrifuge some carrots, we would be left with juice, on the one hand, and a solid, fibrous residue, on the other. Instead of discarding the residue, let's cook it down a bit to concentrate the flavor of its pectins and cellulose, and then, instead of applying an ordinary gel, put this residue in it.

D_3

$D_0 l D_3$: Typically, this assumes the form of a mass of pearls. We have already encountered such a structure in the case of conglomèles. But why not think also of stacked cubes (as in the puzzle Rubik's cube)? Imagine that every other cube is hard, all the intervening ones being soft. The effect in the mouth is incredible, for the teeth, in pressing down on hard and soft textures, transmit signals that the brain interprets as evidence of disequilibrium: first, the sensation of hardness as the teeth bite into the cube on top, then of softness as the teeth reach the cube immediately below, then of hardness again, and so on. This effect can be used in both traditional and note-by-note cooking. One way of rapidly producing such cubes is this: make two flat layers, one hard and the other soft; cut them in strips and then lay them on top of each other in alternation; next, cut the layered mass into squares and then superpose the squares, rotating each new layer by a quarter turn; finally, consolidate the new layered mass by drizzling some sort of liquid cement over it (a solution made with gelatin or some other gelling agent, perhaps a beaten egg, which could then be cooked).

$D_1 l D_3$: Either fibers can be neatly aligned with one another, or they can be thrown any which way into a cubic mold. Obviously, the first option is tidier: one might think of arranging the fibers along the line of their grain or of aligning fibers gathered in bundles, as in the case of meat. (The formula $D_1 l D_3$ describes not only the structure of natural meat, but also the artificial meat I mentioned earlier.) Alternatively, you could mold pasta fibers into a slab and then punch a lattice of holes into the top surface. In traditional cooking, one thinks of the turban of noodles that classically accompanies sole. Or take a piece of Gruyère cheese and carve out a network of canals with a pastry cutter, then drizzle some egg batter into the grooves and cook the whole thing quickly in a microwave oven.

D₂/D₂: A sheet made of several sheets? Isn't that what puff pastry is?

D₃/D₂: Once again, an impossibility.

D₂/D₃: Here again the choice is simple: either you arrange the sheets in an orderly fashion or you don't. In the first case, you might superpose sheets of puff pastry, as in a baklava. In the second case, the sheets might be roughly incorporated in a whipped substance of some sort to form a jellied chiffonnade, for example, or the flaky apple pie known as a *pastis gascon*.

D₃/D₃: Here we are dealing with large consolidated masses: a three-dimensional checkerboard like the Rubik's cube mentioned earlier, for example, but with fewer than ten pieces per dimension—since ten times 2 centimeters fills up the plate.

INCLUSION (SYMBOL @)	D_0	D_1
D_0	D_0 @ D_0: A pearl inside a pearl might be a lovely thing. All of the preceding examples have involved solids, but liquid creations are also possible. For example, one might cover a bit of sorbet with isomalt and then let it thaw. This is more or less the same principle involved in making a croquette.	D_1 @ D_0: Here an object 2 centimeters in diameter must include a one-dimensional object. Think of a strand of spaghetti that has been rolled up into a ball and then held in place by a shell or placed inside a gel, for example.
D_1	D_0 @ D_1: A pearl inside a cylinder, perhaps. Myself, I prefer the D_0/D_1 structure mentioned in table 2.4.	D_1 @ D_1: A fiber of one substance with one of a different substance inside it would make a novel kind of *amuse-bouche* or *mignardise*.
D_2	D_0 @ D_2: A pearl enclosed by a more or less thin slab. This is the principle of ravioli, where a small ball of ground meat is placed between two sheets of pasta that are kneaded together.	D_1 @ D_2: A fiber inside a sheet—a way of breaking the monotony of the sheet, of giving it a direction.
D_3	D_0 @ D_3: A pearl inside a mass. One thinks of Chef Grant Achatz's use of a fresh almond to give visible structure to a cucumber gel. It is very beautiful to behold: the fresh green transparency of the gel sets off the dazzling whiteness of the almond; in the mouth, one senses a correspondingly marvelous contrast.	D_1 @ D_3: A fiber in a volume, creating an axis.

D₂ @ D₀: Take a sheet and then crumple it into a little ball. In traditional cooking, cooked spinach is shaped into a ball, breaded, dipped in egg batter, and fried, for example. In note-by-note cooking, there are many more possibilities, some of which are derived from prior advances in molecular cooking. Begin by dissolving some gelatin in a well-seasoned vinegar, then whisk in some oil. Next, drizzle the emulsion on a marble slab: cooling produces a sheet of vinaigrette. Crumple this sheet and cover it with a salad leaf to make a small object 2 centimeters in diameter—and there you are! This result can easily be extended to accommodate innovations in note-by-note cooking.

D₃ @ D₀: Impossible.

D₂ @ D₁: Here a sheet is rolled up and placed inside a fiber. (It is curious, by the way, that this possibility, like so many of the others inventoried in these tables, has not already been systematically investigated by traditional cooking.)

D₃ @ D₁: Again, an impossibility.

D₂ @ D₂: A sheet inside a thicker sheet. This is not very difficult to do. But let's not limit ourselves to the mere superposition of sheets, for in that case we would fail to grasp the true nature of the system described here.

D₃ @ D₂: This time there is no impossibility, for the sheet can cover the volume. Here we have contrasting sensations of hot and cold: a piece of cooked chicken, for example, dipped in a sauce that forms a jellied surface layer, or aspic, imparting flavor to the meat. Other materializations of the formula are possible, however, from a chocolate candy bar with a soft filling to a small meat pie napped with a liquid sauce.

D₂ @ D₃: A sheet at the center of a volume—of a gel, for example.

D₃ @ D₃: A volume at the center of a volume. Such a structure has already been produced by chefs teaching at Le Cordon Bleu in Paris, who placed a citric acid and glucose sorbet inside a hard red isomalt shell.

SUPERPOSITION

(SYMBOL σ)	D_0	D_1
D_0	$D_0 \sigma D_0$: One pearl on top of another—not much to eat, but it can still be a delicacy. Or perhaps a small cube of caramel surmounted by a small cube of chocolate jelly.	$D_1 \sigma D_0$: A string on a pearl—a nice decorative touch, if nothing else. Perhaps someone can come up with a more interesting idea?
D_1	$D_0 \sigma D_1$: A pearl sitting on a string. It might well be delightful, but probably it would be more pleasing to the eye than to the palate.	$D_1 \sigma D_1$: Four possibilities come to mind: a string on a rope, a string on a cylinder, a rope on a rope, a rope on a string. (It is interesting to note, by the way, that the human eye prefers to see the larger object placed underneath—as in the case of a pyramid, whose base is broader than its apex.)
D_2	$D_0 \sigma D_2$: A pearl placed on a sheet. Pierre Gagnaire, for example, has placed a transparent pearl, its color a superb blood red, on a nasturtium leaf (which itself was placed on top of another essential part of the dish).	$D_1 \sigma D_2$: A pearl on a string could be a lovely thing. There is no reason, by the way, that the string has to be rectilinear. It could have any shape at all.
D_3	$D_0 \sigma D_3$: A pearl placed atop a mass. Gagnaire had the idea of putting a black pearl on a small cushion with red-and-white streaks—though in fact the cushion had the structure $D_3 @ D_2$, with seasoned whipped cream at the center of a paper-thin sheet of raw beef (carpaccio).	$D_1 \sigma D_3$: A string atop a mass. The preceding discussion applies here as well.

$D_2 \sigma D_0$: A sheet on a ball, perhaps a flat piece of pasta (lasagna, for example) laid over a small heap of forcemeat. Keep in mind that the diameter of the forcemeat must be less than 2 centimeters if the sheet covers the plate.

$D_2 \sigma D_1$: A sheet laid over a string (or a rope). This would be a way of calling attention to the string (or the rope) without really showing it. Why should such a thing matter? For the moment, I can do no more than appeal to the necessity of artistic choice—and leave it at that until we come to chapter 6.

$D_2 \sigma D_2$: A sheet on top of a sheet. Isn't that the beginning of a cake?

$D_2 \sigma D_3$: A sheet on a mass—as, for example, the icing on a cake.

$D_3 \sigma D_0$: A slab on a ball? Sure, it could be done. But why bother?

$D_3 \sigma D_1$: A mass on a string or a rope. The immediately preceding point in this column needs to be extended to cover this case.

$D_3 \sigma D_2$: A mass on a sheet. Among other things, this expresses the idea of being seated. A crispy dough covered by a filling, for example. A quiche has exactly this structure.

$D_3 \sigma D_3$: One mass on top of another. This arrangement is found all the time in traditional cooking.

INTERPENETRATION (SYMBOL ×)	D_0	D_1
D_0	$D_0 \times D_0$: This formula makes no sense unless it is reconceived as a version of $D_3 \times D_3$ on a smaller scale (see the bottom-right panel in this table).	$D_1 \times D_0$: Here again the formula is meaningless unless it is somehow pictured as a string wound up inside a "point." Let's not dwell on this—there are many other possibilities to explore.
D_1	$D_0 \times D_1$: Uninteresting.	$D_1 \times D_1$: Impossible.
D_2	$D_0 \times D_2$: Uninteresting.	$D_1 \times D_2$: A string that passes straight through a sheet.
D_3	$D_0 \times D_3$: Impossible since a point does not have the same "size" as a volume.	$D_1 \times D_3$: A string that passes straight through a volume.

$D_2 \times D_0$: This formula is likewise devoid of interest.

$D_3 \times D_0$: Here the formula cannot be given physical expression.

$D_2 \times D_1$: Impossible.

$D_3 \times D_1$: Impossible.

$D_2 \times D_2$: Two continuous interpenetrating lattices inside a "sheet." Isn't this rather similar to the old idea of wrapping a book in marbled endpapers? If the colored ink is replaced by flavored liquids, one may expect to find a little bit of happiness.

$D_3 \times D_2$: Impossible, unless reconfigured as a $D_2 \times D_3$ structure.

$D_2 \times D_3$: A sheet that passes straight through a volume.

$D_3 \times D_3$: This is the case of a cross-linked gel made with gelatin or with any other gel whose solid phase is a lattice. Note that the proportion of solid and liquid phases may differ—hence the considerable variations in consistency. There is a world of difference between a gel with 1 percent gelatin and a gel with 10 percent gelatin.

is a risk that surveying the range of possible combinations will grow tiresome, but there is the chance of reward as well: just as molecular cooking proudly took the siphon bottle and liquid nitrogen as its emblems, so it may be that new consistencies will prove to be the distinguishing marks of note-by-note cooking. Nevertheless we must be careful: just as molecular cooking cannot be reduced to a few simple preparations, so note-by-note cooking cannot be limited to a small class of elementary combinations. A flag is not the same thing as the people it stands for!

Even so, molecular cooking has already shown that progress has followed upon attempts to devise practical recipes. When I proposed using alginates to make liquid pearls, cooks were not so much reluctant to experiment as unfamiliar with the technique involved. Once they understood that they had only to dissolve 5 grams of alginate for every 100 grams of liquid and 5 grams of calcium lactate for every 100 grams of liquid, things went smoothly. The same will be true in the case of note-by-note cooking—which is why I have collected a number of recipes in the appendix to this book. It is also why I beg you to indulge me as I comment, occasionally at some length, on the combinations summarized in tables 2.4 through 2.7.

Neither random dispersion (table 2.4) nor inclusion (table 2.5) has been systematically explored by traditional cooking. There has been some interest in investigating primitive forms of superposition (table 2.6), but only in a rather desultory way. As for the interpenetration (or imbrication) of two continuous networks (table 2.7), a somewhat more complicated exercise from the technical point of view, virtually nothing has been done. All the more useful, then, will it be to have a schematic inventory of possibilities to hand, from which one can readily choose without having to mentally recreate the list again and again, at the risk of omitting one or more of the panels each time.

Off we go, then—bearing in mind, as I say, that however large the number of possible combinations of only two objects may be, many more exist with three objects, still more with four, and so on. In fact, the situation is far more complicated still. For each of the possibilities listed in these tables, nothing whatever is said about the exact shape of the objects assembled: from the dimensional point of view, a square sheet and a disk are identical; a pyramid and a cube are indistinguishable. And there are infinitely many such forms!

Figura, Ludger O., and Arthur A. Teixeira. *Food Physics: Physical Properties—Measurement and Applications*. New York: Springer, 2007.

Foster, K. D., J. M. V. Grigor, J. Ne Cheong, M. J. Y. Yoo, J. E. Bronlund, and P. Morgenstern. "The Role of Oral Processing in Dynamic Sensory Perception." *Journal of Food Science* 76, no. 2 (2011): R49–R61.

Guérard, Michel. *La grande cuisine minceur*. Paris: Robert Laffont, 1976; partly translated in Michel Guérard, *Cuisine for Home Cooks*. New York: Morrow, 1984.

Hargittai, István, and Magdolna Hargittai. *Symmetry Through the Eyes of a Chemist*. 2nd ed. New York: Plenum Press, 1995.

Hubel, David H. *Eye, Brain, and Vision*. New York: Scientific American Library, W. H. Freeman, 1988.

Thibault, Jean-François, and Marie-Christine Ralet. "Pectins: Their Origin, Structure, and Functions." In Barry V. McCleary and Leon Prosky, eds., *Advanced Dietary Fibre Technology*, 369–379. Oxford: Blackwell, 2001.

This, Hervé. "Formal Descriptions for Formulation." *International Journal of Pharmaceutics* 344, nos. 1–2 (2007): 4–8.

——. *Science, technologie, technique [culinaires]: Quelles relations?* Cours de gastronomie moléculaire no. 1. Paris: Belin-Quae, 2009.

THREE ▸ TASTE

NOTE-BY-NOTE COOKING will sound the death knell for the false theory of the four tastes. Once chefs and gourmets taste pure compounds, they will notice real differences that until now have been hidden from them by four simplistic and misleading terms: *salty, sweet, acid, bitter*. Words do sometimes lead us astray, it is true. A new vocabulary will have to be introduced.

The difficulty facing anyone who wishes to have a better understanding of taste is that gustatory perception is influenced to a very substantial degree by color, consistency, smell, and so on. When we eat, our perception of tastes is inevitably mixed with the perception of odors because odorant molecules released by chewing rise up into the nose. It is therefore possible to isolate tastes only by closing off the retronasal channel. You could do this, of course, by holding your nose while eating, but a better way would be to use a small aquarium pump to blow air through a tube inserted in one nostril while pinching the other; alternatively, you could make a small mask out of modeling clay and divide the stream of air coming from the pump into two equal streams passing into the nostrils at the same time. The air from the pump blocks the odorant molecules from reaching the nose, so you can perceive the taste of pure compounds.

BUBBLE COCKTAIL

A cocktail made of water, ethanol (about 35 percent), citric acid, glucose, sucrose, tartaric acid, and artificial blueberry flavoring. The mixture is put in individual soda siphons, placed under pressure using carbon dioxide, and dispensed directly from the siphons into serving glasses. First served at a dinner sponsored by the Quebec Institute for Tourism and Hotels in Montreal, April 11, 2012.

▶ BOUCHÉE ULTRA
Made with coagulated fish proteins, triglycerides (oil can be substituted), amylose and amylopectin (otherwise corn starch), brilliant blue (the blue pigment found in Curaçao, for example), glucose, salt, piperine (alternatively, pepper that has been macerated in oil), and compounds that impart a sensation of freshness (cooks may use what the food industry calls "coolers," for example, or fresh cucumbers macerated in water and oil). Montreal, April 11, 2012.

◣ BOUCHÉE TART
Molecular cooking has sometimes been criticized for its use of "soft" products, in particular gels, but this is due to a misunderstanding. All meat and vegetables are gels, consisting as they do of water trapped in a solid; moreover, traditional and molecular cooks alike can make whatever soft or hard products they need. Here the hard, glassy top part was made from isomalt and placed on top of a red base made from wheat proteins. Montreal, April 11, 2012.

BOUCHÉE NUAGE
This dish—made from gluten, which is not, strictly speaking, a pure compound, but rather a mixture of wheat proteins—is an example of "practical" (rather than "pure") note-by-note cooking. It may be likened to using a synthesizer rather than a computer to make musical sounds. Artificial color, flavor, and odor are added to the gluten once it has been cooked. The glassy, green element is made of isomalt, colored with a green pigment, with flavor and odor added as well. Montreal, April 11, 2012.

LE CORDON BLEU

A dessert consisting of five different layers, prepared by Jean-François Deguignet, chef-instructor at L'École Le Cordon Bleu Paris and served as part of an annual dinner for students of the Institute for Advanced Studies in Gastronomy (IASG) held at the Cordon Bleu. It was made using a blue pigment, of course, but also with a number of other compounds that are officially classified as food additives. The flavor is rather like strawberry. Paris, October 15, 2011.

NOTE-BY-YOLK

This is not an egg dish, however much it may resemble an Oeuf en meurette. The confusion was deliberate and gave rise to an amusing discussion among the guests at the 2011 IASG/Cordon Bleu dinner, who found the sauce to be quite acidic. In fact, the level of acidity is less than that of an ordinary salad dressing, but the guests had anticipated an egg flavor that was not present. Paris, October 15, 2011.

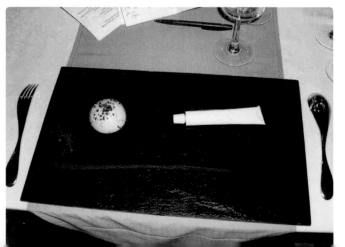

ARDOISE THIS

A dessert first served at the 2010 IASG/Cordon Bleu dinner. The "macaroon" on the left was made using egg proteins, sucrose, and tartaric acid. In the tube on the right, a red emulsion was prepared from cocoa butter, orange powder, and lecithin. Paris, October 16, 2010.

IASG/CORDON BLEU CHEFS
Chef Patrick Terrien and members of the team that prepared
the note-by-note menu for the 2012 IASG/Cordon Bleu
dinner. Paris, October 20, 2012.

SP"READY" TO M"EAT"

Veal tongue fibers were mixed with turnip cellulose and arranged on a spatula with alternating layers of fatty and lean matter. The composition was then solidified with egg ovalbumin, oil, veal stock that had been slowly reduced (to hydrolyze proteins and make amino acids), citric acid, and polyphenols from the juice of Syrah grapes. Paris, October 20, 2012.

⌐ MOUSSE-NOTE "BABYBEL"

This miniature version of a commercially produced French cheese was made from milk powder, yogurt powder, iota- and kappa-carrageenan, and salt. Dipped first in a bath of water, red colorant, and iota-carrageenan, then served on a foam made with milk powder and lecithin. Paris, October 20, 2012.

◣ STRAWBERRY AND LEMON TART REVISITED

The strawberry tart is a traditional dessert, sometimes served with lemon cream. In this version, devised by Chef Jean-François Deguignet, all the ingredients (except the strawberry powder) are pure compounds: lecithin, citric acid, glucose, sucrose, amylopectin, calcium lactate, water, sodium alginate, agar-agar, carrageenan. Paris, October 15, 2011.

RED AMUSE-BOUCHE

Prepared by chefs of the Paris chapter of Les Toques Blanches for the chapter's 2011 Telethon, under the supervision of Chef Jean-Pierre Lepeltier, the dish consists of a salty aqueous solution with red colorant, ethanol, tomato powder, tartaric acid, and glucose. The green leaf on top is made of isomalt. Paris, December 3, 2011.

NOTE-BY-NOTE NO. 1
The first note-by-note dish ever
served in a restaurant, by Pierre Gagnaire. The bottom part consists of alginate pearls containing a mixture of
water, ethanol, salt, glucose, glycerol,
and tartaric acid. It lies beneath a series of five glucose péligot disks (similar to caramel), topped by sherbet.
Hong Kong, April 24, 2009.

EMULSION OF SEPTEMBER 20
A dessert made by the chefs Jean-
Pierre Lepeltier (Hôtel Renaissance
Paris La Défense), Michaël Foubert
(Hôtel Renaissance Paris Arc de
Triomphe), Lucille Bouche (Hôtel
Renaissance Paris Trocadéro), and
Laurent Renouf (Hôtel Renaissance
Paris La Défense) for the dinner
held at AgroParisTech in September
2012 to mark the publication of the
French edition of this book. An actual
chicken bone is encased in a mixture
including proteins, ethanol, water, oil,
tomato extract, and sucrose. Paris,
September 12, 2012.

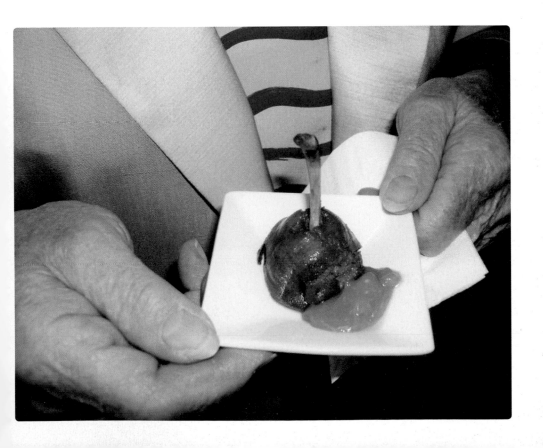

JAMBONNETTE
A dish conceived by the chefs Julien Lasry (Hôtel Renaissance Paris La Défense) and Jean-Pierre Lepeltier. The "bun" of these minihamburgers is made from amylopectin, ovalbumin, salt, and water. The "meat" is made from water, egg proteins (but any coagulating protein could be used), cellulose, and a number of flavoring ingredients. Paris, September 20, 2012.

ORANGE BALLS OF FIRE

A fiery appetizer created by Chef Laurent Renouf consisting of orange-colored flattened foam spheres, set alight with a mixture of ethanol and water. Served as a flaming dessert, it includes sucrose. Paris, September 20, 2012.

AMYLOPECTIN TARTS

First served by Chef Lucille Bouche, these tarts (bouchées) were made by cooking a mixture of roasted amylopectin, ovalbumin, and water in triglycerides (otherwise in oil) for 10 minutes. The soft white "ball" placed on top is made from yogurt powder, milk powder, iota-carrageenan, and agar-agar. Paris, September 20, 2012.

CHICKEN PEARLS

Many of the gelling agents first introduced by molecular cooking (in this case sodium alginate) can be used in note-by-note cooking as well. Most traditional food ingredients are gels—plant or animal tissues in which water is trapped in a solid. The dish shown here was served as part of the 2012 IASG/Cordon Bleu dinner. Paris, October 20, 2012.

ORANGE CONGLOMÈLE

Natural oranges are made of many small vesicles (or sacs) of juice "glued" together and enclosed by a skin. Here we have small artificial pearls made from water, citric acid, and sugar and enclosed by a skin obtained by gelling sodium alginate with calcium salts. In this case, the procedure for making pearls was reversed: calcium was dissolved in the artificial juice, and then sodium alginate was dissolved in water to which this juice had been added, drop by drop; next, the pearls were collected and "glued" together using gelatin in which limonene and citral had been dispersed. I refer to all such systems as "conglomèles."

FIBRÉ

Artificial meat can be obtained by growing living cells, but the fact that meat consists of fibers connected by collagenous tissue suggests another method. The hollow tubes pictured here are made of fibers containing water and proteins and connected to one another by collagen molecules dispersed in water, just as in animal muscle tissue. I call all such examples of artificial meat "fibrés."

MISDIRECTION AND MISPERCEPTION

The gourmet who is serious about exploring the world of tastes will also have to dissolve sapid compounds. Why? Because otherwise he will be fooled by the consistency of the foods he eats (as you can see for yourself by comparing the taste of sea salt, table salt, and salt that has been finely ground). Color will have to be disguised as well (sensory analysis is often done by putting subjects in rooms lit with red light, but you can also simply close your eyes). Having taken these precautions regarding smell, consistency, and sight, begin by tasting whatever happens to be in your pantry: wine, chocolate, the oil in which sardines are packed, Roquefort—you name it. Just as novice oenophiles must learn to taste wines, aspiring gourmets will have to be taught to recognize the different tastes of the foods they eat, starting with traditional dishes and moving on to pure sapid compounds and artificial flavoring substances. Keep in mind that compounds can be both sapid and odorant. It is true that salt and sugar have no smell when they are pure, but what about the ethanol in brandies? It has a smell, quite obviously, but it also has a taste. Moreover, the taste depends on its concentration when it is mixed with water. This is a phenomenon that sensory physiologists are well familiar with: even the taste of salt and sugar varies according to their concentration; indeed, at very low concentrations in water, salt can actually taste sweet.

Note, too, that if the theory of the four tastes is false, the familiar map of taste-receptor areas on the tongue is no less mistaken, no matter how many times it may have been reproduced in monographs, textbooks, and works of popular science. It is a mystery that this map should still have any credibility at all. One has only to put a few drops of sapid solution on different parts of the tongues of different people to see that the map changes from person to person. It is customarily said, for example, that the tip of the tongue recognizes sweetness—a doubtful notion to begin with since there are several kinds of sweetness. As it happens, many people sense sweetness less on the tip of the tongue than on other parts of the organ. (Personally, I register a variety of sour sensations there.) When participants in a seminar held in Paris a few years ago were invited to stick their tongues in a glass of sweetened water, only 40 percent of them sensed a sweet taste in the tip of the tongue.

How do such misconceptions gain currency? The answer seems to be that human beings are susceptible to arguments from authority: we willingly adopt the ideas of acknowledged experts, no matter how far-fetched they may be. We are, after all, ourselves primates, a group of animals in which certain dominant individuals are able to persuade the other members of the group to do more than they could be made to do by force alone. Nevertheless it is also true that in matters of taste it is what each of us likes personally that we consider to be good: if we don't like a dish, we can't imagine it's worth eating—even if we know that a billion people are wild about it; conversely, if we like something, there's no point arguing about it, unless of course we want to argue with friends for the sake of cultivating a spirit of open discussion and strengthening social ties. This last point seems to me essential. When we debate the meaning of existence over a meal in the corner bistro, we do so as part of a group. Being part of a group is the crucial fact of life for members of a social species.

But I digress. Anyone who wishes to construct note-by-note dishes will have to acquaint himself or herself with the tastes of compounds in order, as I have already said more than once, to learn to say new words in a new language and then to make new sentences. No doubt this is why some cooks will resist note-by-note cooking. Understandably, they do not relish the thought of being dispossessed of everything they have learned until now—of being stripped naked, as it were, both intellectually and professionally. But this will be true only for cooks who are timid or lazy; for the others—the artists, the innovators, the brave and industrious ones—the prospect of discovering a new world will be experienced as an immense thrill and welcomed as an incomparable opportunity.

THE IMPOSSIBLE DESCRIPTION OF UNKNOWN TASTES

Earlier we saw that although shapes can be readily enough described, consistencies are harder to imagine when we have not encountered them before. What about tastes? Suppose you had never tasted sweetness. We do, of course, have certain points of reference for naturally occurring compounds in the vegetable and animal kingdoms: the tastes of meats, fruits, and vegetables that contain these compounds in significant concentrations. Anyone who has tasted sugar cane or honey can easily imagine the taste of sucrose.

Even so, the ability to construct note-by-note dishes requires a period of apprenticeship. As in the case of consistencies, the wisest course will be to acquaint ourselves with the taste of pure sapid compounds before sampling pairs of such compounds, triplets, and so on. By itself, however, tasting isn't enough, for being human also means sharing. This is why we must also devise a new vocabulary of tastes that goes beyond the worn-out mantra "salty, sweet, sour, bitter," which for many years now has been refuted both by personal experience and by sensory physiology. Consider, for example, sodium bicarbonate, which is used in cooking chiefly to soften dry vegetables and to keep green vegetables green. It has a somewhat soapy taste, but not quite soapy either; it is sweetish, but not really sweet; slightly salty, but at the same time something else. How are we to describe it? Obviously we could speak of "bicarbonate of soda taste," but this would force us to introduce as many such terms as there are compounds! For the glycyrrhizic acid of licorice, we would then have to say "glycyrrhizic acid taste"; for glucose, "glucose taste"; for aspartame, "aspartame taste"; for tyrosine, "tyrosine taste"; and so on without end.

This difficulty is compounded by the fact that human beings perceive different tastes differently. An excellent example is monosodium glutamate, much used by cooks in parts of East Asia and by certain sectors of the food industry in Western countries. It is derived from glutamic acid, which, together with other amino acids, becomes linked in chains of various lengths to form proteins, and in particular the very proteins on which the functioning of the human body depends. Like all acids, it loses a hydrogen atom when it is dissolved in water, forming a glutamate ion. If this glutamate ion is combined with sodium ions (sodium atoms that have lost an electron), one obtains monosodium glutamate in the form of small, white, needlelike crystals.

When a fairly large number of people chosen at random taste monosodium glutamate, they fall into three groups: some subjects perceive a salty taste, some a sweet (or at least a very mild) taste, and still others an assertive taste of chicken broth. These sensations seem to correspond to individual genetic differences. Moreover, the phenomenon is almost surely not re-stricted to the perception of tastes. After all, it is not because we call a certain color blue that all of us perceive it in the same fashion. This suggests that it will be possible to find a common gustatory language if we start from a shared set of conventions.

SAPID COMPOUNDS

What is there to be said, then, about sapid compounds? To be honest, we know scarcely anything at all about them. It is true that I shall go on in a moment to examine a considerable variety of sapid compounds about which a great deal is known. But much more remains yet to be discovered before we can claim to have a proper understanding of this vast and so far largely uncharted world.

Let us begin this time by adopting a biological point of view. Why do we perceive tastes? The question is poorly put because yet again it seems to sug-gest that biological evolution has a purpose. Few people really believe that human beings were endowed with the faculty of perceiving tastes, which I have proposed to call "sapiction," so that they could take pleasure from eat-ing. No less than our ancestors, we are animals, and if we are alive today it is only because our ancestors succeeded in reproducing themselves. And if they were able to reproduce, it is because they managed to reach sexual maturity, which means they found a way to feed themselves. And if they were able to feed themselves, it is because their sapictive system allowed them to recog-nize foods, which is to say edible things from which they could absorb nour-ishment and energy. As the anthropologist Claude Marcel Hladik has empha-sized, nonhuman primates (the "apes") coevolved with plants that produced sweet fruits. Our ancestors benefited from ingesting the sugars of these fruits and in dispersing their seeds and pits caused sweet fruits to be widely prop-agated. The consumption of toxic plants, by contrast, was discouraged by the presence of various compounds, typically bitter in taste, with the result that their propagation depended on other mechanisms of dispersal. Still today, human beings commonly consider bitterness to signal danger. But plainly this is not always true. Beer is perfectly safe to drink, even though the alpha-humulene contributed by hops makes it bitter.

It should be clear that opposing sweetness to bitterness serves no real purpose. The only things that matter are the specific proteins, known as receptors, on the surface of the papillary cells in the mouth and on the tongue that register specific tastes. When we eat a food, it releases various compounds; some of these compounds dissolve in saliva, as we have seen, and some of those that dissolve can bind with a receptor, producing a sapid sensation. Although water solubility is a necessary condition of our perceiving taste, it is not a sufficient one: some compounds that dissolve in saliva have no taste, quite simply because neither the mouth nor the tongue contains the receptors needed to detect them.

The word *detect* is really a bit too strong, for it is a verb of action. Taste receptors consist of inert strings of protein that are more or less compactly folded up on the surface of the papillary cells. Owing to the random motion of molecules in solution, one such molecule may pass in the vicinity of a receptor. A sapictive sensation is registered when a sort of complementarity between the molecule and the receptor allows a bond to be established. Think of the receptor as a string on which north and south magnetic poles are located at various places: if the compound that comes near to a receptor is complementary, then the molecule–receptor connection is made; if not, nothing happens. In those cases where a compound binds with a receptor, a series of molecular transformations causes an electrical signal to be transmitted from the cell bearing the receptor to the brain. There, in the brain, the signal is interpreted.

A compound may therefore be said to be sapid if it binds to a receptor. But a taste can also appear if two small molecules, neither of which is capable by itself of stimulating a sapictive sensation, succeed in binding to two parts of the same receptor. There may be any number of other possible configurations as well. As I say, the world of sensory physiology is still only very incompletely understood. Most of us would agree, I think, that enlightened food lovers should help to make the case for public funding of further research on this fundamental aspect of human experience. For the time being, however,

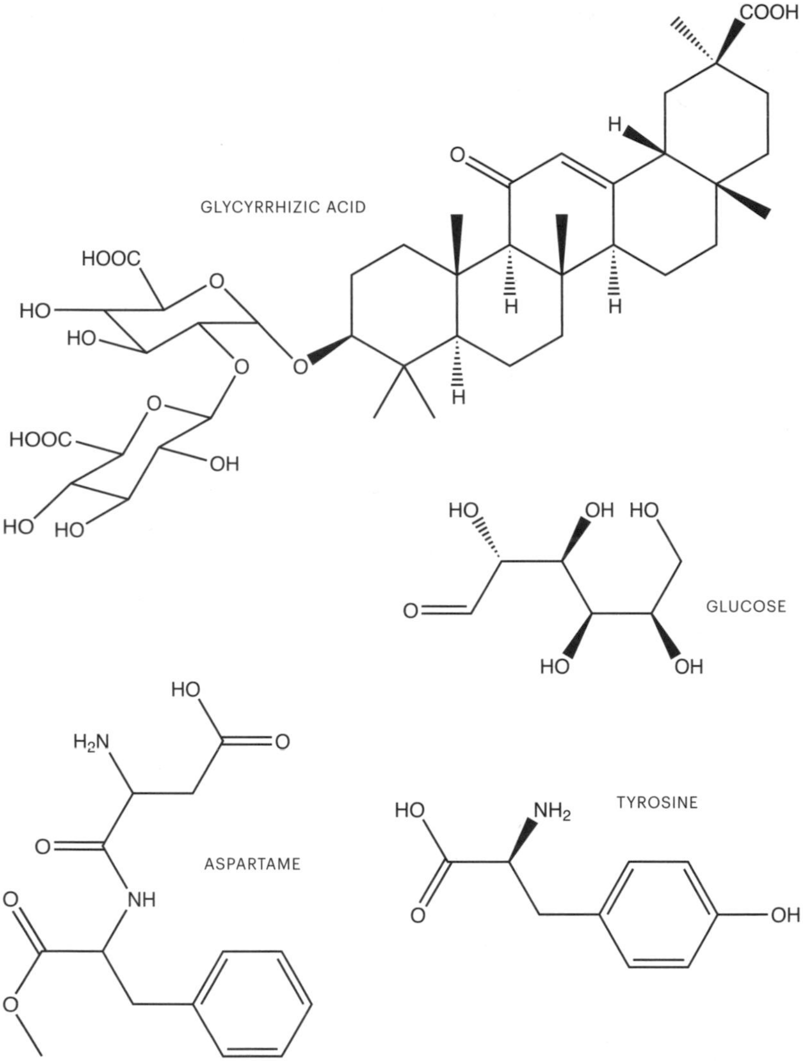

GLYCYRRHIZIC ACID

GLUCOSE

ASPARTAME

TYROSINE

3.1 SAPID COMPOUNDS BIND WITH RECEPTORS IN THE MOUTH, CAUSING ELECTRICAL SIGNALS TO BE SENT TO THE BRAIN.

let's make do with the knowledge we have and take a quick look at some of the sapid compounds that we already understand fairly well, organizing them into a number of general categories for the sake of convenience.

MINERAL SALTS

The term *mineral* here denotes compounds that are not organic. Organic compounds are so called because they were found first in living organisms. The term is unfortunate, however, for organic compounds are now known to exist in everything and everywhere, even in interstellar space. Chemists used to suppose, following the Bible and the Quran, that the world is divided into two separate kingdoms, the living and the inanimate. The inanimate (or mineral) kingdom could be explored, and it was. In the course of examining various minerals, especially ores, chemists noticed that heating some of them produced acids and bases (or alkalis). The calcination of carbonates, for example, brings about the formation of a caustic substance, quicklime, which in water becomes slaked lime, an alkali with a mild taste. (But don't dare ingest it in too concentrated a form!) Similarly, macerating wood ashes in water yields a caustic aqueous solution known as potassium hydroxide (caustic potash or lye).

It was soon observed that acids react with bases and form less corrosive substances, called mineral salts. Thus hydrochloric acid reacts with caustic soda (sodium hydroxide) to produce sodium chloride—which is nothing other than table salt. In reaction with potash, the same hydrochloric acid forms another mineral salt, potassium chloride. Starting from mineral substances, in other words, we obtain a class of substances that are neither acids nor bases.

In the early days of modern chemistry, these substances seemed very different from the organic compounds associated with living organisms. It was gradually discovered that the molecules of these compounds are typically composed of carbon, hydrogen, oxygen, and nitrogen atoms, sometimes in the company of sulfur and a few other elements. It came as a complete shock, then, a seismic upheaval, when the German chemist Friedrich Wöhler (1800–1882) discovered that heating a mineral substance could produce an organic

3.2 *THE GERMAN CHEMIST FRIEDRICH WÖHLER.*

compound. The barrier between the living and inanimate worlds suddenly collapsed—a terrible blow to the doctrine known as vitalism, according to which the living world contained something more than the inanimate world (a divine spark, as it was usually conceived). Now that biologists are able to construct viruses from scratch, synthesized compound by synthesized compound, only the ignorant can go on being vitalists—unless, of course, vitalism today is not a matter of ignorance, but of faith. In the latter case, not even the long-anticipated synthesis of a living cell (now an accomplished fact), molecule by molecule, will be enough to do away with vitalism, which will have taken refuge in objects other than physical bodies, substances, minerals, and atoms.

Let us come back, then, to more down-to-earth questions. Where are mineral salts found? In water, to begin with. On the label of every bottle of mineral water, there appears a list of ingredients: chloride, sodium, calcium, phosphate, hydrogen carbonate, nitrate, magnesium, fluoride. These ions are what give water its taste. Evidently our teachers in school were not telling the truth when they told us that water has no taste, for it is well known that no two mineral waters taste the same. Their taste depends on the proportion of mineral salts they contain. How can the salts be separated from the water? Evaporating the water is one way, though not a costless one (assuming you use heat) or a very entertaining one (waiting for the water to boil off). The ions will eventually end up being deposited at the bottom of the pan, but in only very small amounts: the residue is seldom as much as a gram per liter of water; typically, it is on the order of 0.1 gram per liter. Another, more efficient method is to use one of several modern filtration systems, which operate on the same principle as the removable filter found in the water-filtration pitchers sold today in grocery stores. The recovered dry residue will in any case consist of a mixture of calcium, sodium, fluoride, carbonate, and other ions, and, as a practical matter, you won't be able to taste any of them individually.

Indeed, experiencing the particular taste of an ion is impossible in principle, for ions are either positive or negative. An ion is an atom or a group of atoms that has lost or gained electrons, becoming a very powerful sort of magnet in the process (electromagnetic rather than magnetic, however). This means that cooks will never succeed in isolating ions of a single kind, unmixed with ions having an opposite electrical charge. In sodium chloride (table salt), for example, whether it is rock salt or sea salt, there will always be as many chloride ions as sodium ions; in potassium nitrate (known also as saltpeter, used in curing meats), there will always be as many potassium ions as nitrate ions, which are composed of one nitrogen atom and three oxygen atoms. The pure taste of an individual ion, as opposed to the taste of mixtures of ions, will therefore forever remain beyond our grasp (unless perhaps in the case of organic acids, but that's another story). Even so, let's not give up quite yet. It may yet be possible to have at least some idea of what ions taste like. Get hold of some mineral salts (they're ridiculously cheap, sold by the ton), dissolve them in water in low concentrations, and taste each solution separately; then mix them together, taking care to ensure that the overall concentration in these "cocktails" does not exceed the levels typically found in bottled mineral water—and that none of the ions is poisonous.

It is well known, for example, that there are ignorant (perhaps criminally ignorant) winemakers who treat their casks with volcanic sulfur. Any chemist can tell you that such "native" sulfur is liable to be contaminated by arsenic, an element that when burned produces arsenic oxide (a substance that used to be reserved for mothers-in-law). It is well known, too, that the Romans sweetened their wines with lead salts, especially lead acetate, a water-soluble salt. Some historians now wonder whether this practice may not ultimately have been responsible for the fall of the empire (lead is extremely toxic and would have poisoned the elites who drank such wines). Copper is scarcely a more prudent choice. You should think twice about spraying your garden with copper sulfate (following the example of winegrowers in France and elsewhere, who belatedly stopped spraying their vineyards with it after years of intensive treatment). We will do better, at least starting out, to limit ourselves to the ions found in mineral water. The problem is that although elements such as aluminum, scandium, titanium, chromium, vanadium, cobalt, and zinc do figure among the essential trace elements (oligoelements) studied by nutritionists, they occur in food only in very small quantities. Even talking about grams is out of the question! Selenium, for example, is found in fish in a proportion of 30 micrograms (a microgram is a millionth of a gram) per 100 grams of flesh and in meat in a proportion of 6 to 10 micrograms per 100 grams. An observed proportion of 55 micrograms per liter in the blood is sufficient to establish the presence of cardiovascular risk.

Again, as a practical matter, what does a microgram mean? A milligram? A gram? Georges-Auguste Escoffier's *Le Guide culinaire* (1903), venerated in some quarters still today, courted ridicule by instructing cooks to season two eggs with 32 centigrams of salt, or 0.32 gram—an astonishing degree of precision at a time when kitchen scales measured to the gram, at best. One imagines that Escoffier (or perhaps one of his coauthors, Philéas Gilbert or Émile Fetu) obtained this value by dividing some measurable quantity of salt by the number of eggs called for in a recipe. Assuming the number to be six, rather than two, yields a quantity of 1 gram, just about. But dividing 1 by 3 yields 0.333333 ad infinitum, not 0.32. To specify an exact number of centigrams in this case implies, altogether falsely, that gourmets are able to tell the difference between 0.3 and 0.4 grams of salt. Escoffier should have said instead:

"Season in a proportion of roughly one-third of a gram of salt per egg." That would have been more in keeping with customary practice. The moral of the tale? Undue precision is not only useless, it's absurd!

And how much is a gram, really? Using a precisely calibrated scale to weigh the contents of a tablespoon of salt or sugar, you will usually get values between 5 and 10 cubic centimeters (a cubic centimeter is the volume of a cube each of whose sides measures one centimeter). The contents of a teaspoon are generally reckoned to be between 2.5 and 3.5 cubic centimeters—a figure that reflects not only minor variations in the actual size of different teaspoons, but also in the levels to which the spoons are filled. I had the idea once of pouring granulated sugar into the same teaspoon four times in succession and weighing the result. The values I got were 3.8985 grams, 4.1201 grams, 3.5644 grams, and 3.8892 grams. In other words, any given teaspoon contained about four grams of granulated sugar, but the exact value varied considerably from one teaspoon to another.

The time has now come to exchange theory for practice. Probably not many have actually tasted mineral salts. I have. Jacques Decoret in Vichy has experimented with salts recovered from the spring waters of his native region, which makes him the first chef I know of to use mineral salts in his cooking. The chefs most likely to follow him are ones with close ties to the world of winemaking, where aluminum salts are used when the fermentation process is in danger of being arrested because there are no more nitrogen compounds left for the yeast to absorb. The principal products authorized for winemaking in Europe are ammonium sulfate and ammonium phosphate, with a maximum approved dose of 0.3 grams per liter; ammonium sulfide and ammonium hydrogen sulfide are also permitted, though not in a proportion exceeding 0.2 grams per liter. These substances can't help but change the taste of wine!

Salts are also authorized as additives in wine production, especially for purposes of preservation. A moment ago I mentioned the practice of sulfurizing wine barrels. This used to be done by burning sulfur to form sulfur dioxide, a substance that is liked neither by microorganisms (which are killed by it) nor by human beings (who are apt to get terrible headaches from it). Sulfides (E221–228) and sulfurous anhydride (another name for sulfur dioxide,

E220) need to administered with caution. Among food preservatives, saltpeter has long been used to lengthen the shelf life of hard sausages and hams.
Today the list of authorized additives includes both nitrates and sodium and
potassium nitrites (E249–252), as well as boric acid (E284) and sodium tetra-
borate (borax, E285), which are to be used sparingly. In this connection (recalling yet again the Romans' habit of sweetening wines with lead salts), one
shouldn't suppose that what tastes good is always good for one's health. Nor
should the mostly unproblematic use of such compounds as preservatives
lead anyone to suppose that they can be used in cooking without formal instruction. Chefs must be taught how to handle such products safely, just as
they had to be trained in the proper use of liquid nitrogen in molecular cooking—or, for that matter, in the proper use of knives. All these things pose risks
that must be minimized as far as possible.

ORGANIC AND MINERAL ACIDS

Let's move on to acids, only this time starting with acidifiers, a well-tested
class of derivative substances that appear in the list of approved food additives. These compounds—citric, acetic, tartaric, phosphoric, ascorbic, lactic, malic, and succinic acids, among others—enhance the taste of foods and
beverages. Like their sodium, potassium, and calcium salts, they lower pH
(a measure of acidity, as noted earlier, ranging from 0, for the most powerful
acids, to 14, for the strongest bases).

Chemists know very well that acids are not interchangeable and that acidity is not the only thing that determines their taste (contrary to what the false
theory of the four tastes would seem to suggest). Not only do they have different tastes, but the sensation of acidity is not equally long lasting in every
case. Chemists also know that the addition of an acid makes it possible to
sharpen or soften the perception of odorant compounds. This is why beverage manufacturers delight in exploiting the contrast between sweetness and
tanginess (due mostly to the acidity of phosphoric acid)—a trick they picked
up from traditional cooking. In France, cooks have long used *gastriques*,
sweet-and-sour sauces that typically combine vinegar and sugar; they have
also long been accustomed to add a touch of citrus juice to veal stews just

before serving or a pinch of sugar to wine sauces or a dash of dry white wine to sauces for fish.

Then there is the class of acids proper. Among mineral acids, I mentioned hydrochloric acid (E507) earlier, but there are others: phosphoric acid (E338), sulfuric acid (E513), and so on. Their very names frighten the culinary world, but in low concentrations they are only mildly acidic and present no risk. To put matters in perspective, recall that raspberries are very acidic (they can have a pH as low as 2), whereas diluted hydrochloric acid can have a far higher pH (and therefore far lower acidity). Among organic acids, acetic acid (E260) is one of the best known because it is a product of alcoholic fermentation. It is the principal acid in vinegar, for example, in which it is produced by *Mycoderma aceti*, a microorganism that transforms the ethanol obtained from the fermentation of sugars into acetic acid. In its pure form, acetic acid is called glacial ethanoic acid—a transparent, colorless liquid with an unpleasantly pungent odor that dissolves in water in all proportions. Is it something we should be afraid of? No, because anyone who sniffs glacial ethanoic acid will at once realize that it is not to be swallowed, and because once dissolved in water it turns into something that is both harmless and very familiar—white vinegar! Acetic acid is officially approved today for use as a preservative. This will come as no surprise to cooks, who have long been in the habit of preserving little gherkins (known in France as *cornichons*) in white vinegar.

Lactic acid (E270) has an equally ancient pedigree. Produced by bacterial fermentation of lactose, a sugar found in milk, lactic acid is what gives yogurt and some cheeses their slightly sour taste. It is also found in hard sausages, sauerkraut, pickled cucumbers, turnips, green olives, and so on. In its pure state it is a viscous, nonvolatile liquid. It is authorized both in this form and in that of its sodium salt for certain preparations of black olives as well as in the manufacture of certain new processed cheese products.

Citric acid (E330) is the principal acid in lemon juice. In its pure form, it is a white powder. Many chemists fondly remember making "artificial" lemonades as children by mixing pure citric acid with sodium bicarbonate. Combine the two powders in roughly equal proportions (I myself recommend putting more citric acid than bicarbonate), add water, and voilà, you've got

3.3 *LOUIS PASTEUR STUDIED TARTARIC ACID, WHICH FORMS SALTS AT THE BOTTOM OF WINE BARRELS, AND DISCOVERED THAT CERTAIN COMPOUNDS, LIKE A HUMAN HAND, ARE "CHIRAL"— THAT IS, NOT SUPERPOSABLE ON THEIR MIRROR IMAGE.*

a sparkling and delightfully lemony soda—sweetened with a dash of sugar if you like!

Tartaric acid (E334) is famous for having made Louis Pasteur's reputation. The marvelous advances in microbiology due to this son of a wine-grower (whose vineyard in the Jura is the property of the French Academy of Sciences today) have caused us to forget that he made an extraordinary discovery while studying tartaric acid in grapes. Pasteur was the first to realize that there exist pairs of compounds whose properties differ even though their atoms are identical. Just as a right hand is the image of a left hand in a mirror, so the atoms of these pairs of compounds are ordered in such a way that they form molecules having mirror-image forms; and just as a left hand won't fit into a right-hand glove, so a "left-hand" molecule cannot bind with a "right-hand" receptor. Ignorance of this phenomenon was responsible for the catastrophe of thalidomide, a drug given to pregnant women that caused missing or malformed limbs in their children. The organization of atoms in space is important for odorant molecules (we will see in chapter 4 that limonene, depending on whether it is a right-hand or left-hand molecule, has a smell of lemon or of orange), but also for some sapid molecules as well.

The tartaric acid found in grapes occurs in a form denoted by the symbol L(+)—it would take too long and be of too little interest for our purposes here to explain why—and it is in this form that it is used to acidify liquids and wines. Oenologists and winemakers habitually use tartaric acid in alcoholic fermentation, though getting the proportions right takes practice: if it is added to a wine before the process of cold stabilization (which causes potassium tartrate to crystallize) is initiated, predicting the degree of acidification is difficult. The legal limit is 1.5 grams of tartaric acid per liter for musts and 2.5 grams per liter for wines—roughly a teaspoon per liter of liquid, in other words. Its taste? I leave it to you to judge. Personally, I find it elegantly acidic, unlike acetic acid, which has a rather sharper edge. It would be a welcome thing, by the way, if, in addition to salt and pepper shakers, we were to put a tartrate shaker on our dining room tables. But what, I wonder, should we call a shaker that contains lactic acid?

Perhaps we should stop here before we find ourselves too far afield. Once again, a whole volume could be devoted to acids. We haven't yet even touched on malic acid (from the Latin word for apple, *malum*), ascorbic acid, vitamin C, or, indeed, many others. There's an acid for every taste!

AMINO ACIDS AND THEIR DERIVATIVES

Alongside the particular acids I have already mentioned, there is another set of organic acids having quite different tastes: the amino acids. These compounds are molecules that contain both a group of atoms known as a carboxyl (—COOH) group, present in carboxylic acids (consisting of one carbon atom, two oxygen atoms, and one hydrogen atom) and an amino (—NH₂) group (consisting of one nitrogen atom and two hydrogen atoms). The arrangement indicated by the general formula is never actually encountered in the

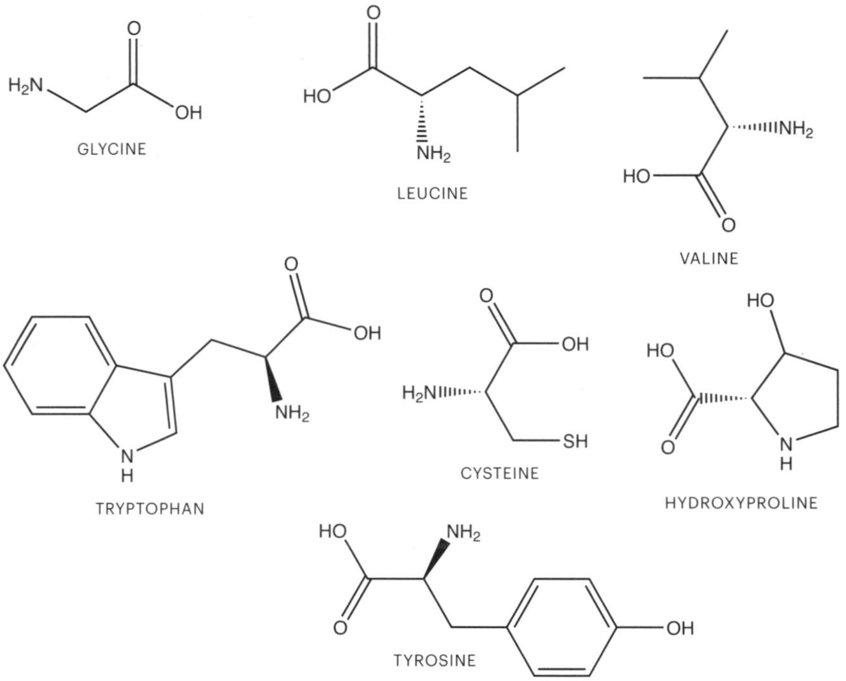

GLYCINE

LEUCINE

VALINE

TRYPTOPHAN

CYSTEINE

HYDROXYPROLINE

TYROSINE

3.4 AMINO ACIDS—GLYCINE, LEUCINE, VALINE, TRYPTOPHAN, CYSTEINE, HYDROXYPROLINE, AND TYROSINE.

real world because, depending on the acidity of a particular environment, the carboxyl group is liable to lose its hydrogen atom or the amino group is liable to gain a hydrogen atom. The molecular details are of little importance for someone just beginning to explore note-by-note cooking without any training in the marvelous science of chemistry. Novices will do well simply to keep in mind that what is written in many textbooks on food science and technology is false—namely, that amino acids are either sweet or bitter.

First, where do amino acids come from? In the human body, they are the links in the molecular chains that form proteins. Proteins are of two kinds: some are what we are made of (the collagen of collagenous tissue, for example, found in muscles and tendons); others, known as enzymes, perform specialized tasks (proteases, for example, serve to cut up other proteins into small pieces, which are none other than amino acids). And so, yes, we are literally full of amino acids. They have various functions in the human body, just as they do in fruits, vegetables, meats, and fish. Proteins, which occur in all living things, are made up of twenty-one amino acids that bear charming names: glycine, alanine, valine, leucine, isoleucine, serine, threonine, cysteine, methionine, aspartic acid, glutamic acid, lysine, arginine, histidine, phenylalanine, tyrosine, tryptophan, proline, hydroxyproline, asparagine, glutamine. What do they taste like? Here again I would have a very hard time describing such sensations in words, but you can try tasting these substances for yourself (health food stores and even supermarkets now sell them). Earlier I mentioned monosodium glutamate, the sodium salt of glutamic acid, which is an amino acid.

All the compounds I discuss in this book can be used to make foods, of course, but there is no reason why they themselves should not be cooked as well. They can be pan fried or grilled or roasted or fermented or braised. The slow cooking of proteins in water, for example, produces a reaction known as hydrolysis, which cuts the proteins up into amino acids as well as fragments called peptides. It is for this reason that gelatin, if it is cooked long enough, no longer sets the cooking liquid in the form of a gel but instead strengthens and intensifies the liquid's flavor. It is for this reason, too, that meats that have

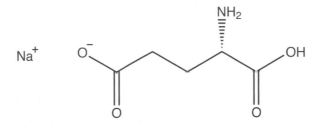

3.5 GLUTAMIC ACID AND THE MONOSODIC SALT OF THIS ACID (BETTER KNOWN AS MONOSODIUM GLUTAMATE).

been slowly braised acquire a marvelous flavor that makes you want to keep on eating. And it is also for this reason that the fish sauces of Southeast Asian cuisines, not unlike the *garum* of the ancient Romans, impart so much flavor: monosodium glutamate is neither a great culinary novelty nor a Japanese invention; cooks in many countries have been making preparations of this type since antiquity, widely imitated thereafter.

Allow me, if you will, to anticipate somewhat the discussion of sugars that follows and mention a class of reactions named after the French chemist Louis Camille Maillard (1878–1936). Maillard discovered that amino acids react with so-called reducing sugars, such as glucose, with the result that, following an initial condensation reaction (in which an amino acid molecule binds with a sugar molecule and hydrogen and oxygen are eliminated in the form of water molecules), rearrangements of various atoms produce a great many odorant, sapid, and coloring compounds. Although these Maillard reactions are not solely responsible for the taste of bread crusts and roasted meats, they do decisively contribute to it. Amino acids can nevertheless react and be transformed even if they are heated in water without sugar. Cooking cysteine, for example, releases a gas called hydrogen sulfide. In large quantities, this gas is both nauseating (it has a smell of rotten eggs) and toxic; indeed, it is all the more dangerous as the first sign of intoxication is masked by the fact that, because hydrogen sulfide deadens the sense of smell, its presence can no longer be detected. In lesser amounts, however, it imparts an agreeable taste of cooked egg.

These few examples will suffice, I trust, to show that the cooking of amino acids may form the basis of a theory of new compounds, just as the cooking of sugars to make caramels inspired research on the chemistry of sugars more than a century ago—which leads directly to my next topic.

SUGARS

Normally, a lengthy introduction to this sizeable group of compounds would be required, distinguishing between carbohydrates, saccharides, monosaccharides, and so on. But it would serve no very useful purpose in this case, and I propose to dispense with it altogether, except for a few preliminary words regarding the term *carbon hydrate*. A little historical background is needed to understand why chemists consider it to be a misnomer. Modern chemistry began with elemental analysis: for any given pure compound, chemists sought to determine the quantity of carbon, hydrogen, oxygen, and other elements it contained. In the case of sucrose, for example, they found twelve parts carbon for twenty-four of hydrogen and twelve of oxygen. Having detected twice as much hydrogen as oxygen, as in the case of water, they concluded that sucrose contains water in addition to carbon—more precisely, that it contains one part water for every part carbon. This should not seem so surprising, really, for chemistry was still in thrall to the ancient doctrine of the four elements, according to which all matter is made up of earth, fire, air, and water. In short, imagining that sucrose consists of water and carbon, chemists considered sucrose to be a carbon hydrate, just like the other sugars, which, as it happens, contain the same proportion of one part carbon for one part oxygen and two parts hydrogen.

Elemental analysis nevertheless said nothing about the structure of molecules, and chemists soon came to realize that they were dealing instead with a family of so-called polyhydroxylated compounds, in which several carbon atoms bear a hydroxyl group consisting of an oxygen atom joined to a hydrogen atom. I won't go into the details, which chemists find fascinating, but which most note-by-note cooks and general readers will not need to know. It is enough to say that there exist on the one hand simple sugars and on the other more elaborate sugars formed by linking simple sugars together to form chains, just as proteins are formed by creating chains of amino acids. There is, however, one essential difference: whereas proteins are linked in linear chains only, sugars may be linked in both linear and branching chains.

THE SIMPLEST SUGARS

The simplest sugar is glycerol, also known as glycerine (E422). It is used in making wines and has a mild taste. It is still seldom used in cooking, but that may change before long.

Glucose is only slightly more complicated. Present in most vegetables and abundant in carrots, onions, fruits, and honey, it serves, as we saw earlier, as a source of fuel for the human body. Glucose is not officially classified as an additive, though it can be used as one. In its pure state, it is not a syrup, as pastry chefs (who buy it in this form) are inclined to believe, but a white powder with a singular taste. Its appeal can readily be seen by performing a simple experiment: add glucose to any classic sauce, and you will see that it enhances the taste in a most remarkable way. What is more, it is very easy to prepare: all you have to do is hydrolyze a modified starch (as the food industry has done for many years now in manufacturing the glucose syrup generally used in pastry making).

Fructose, like glucose, is found in carrots, onions, and other vegetables, as well as in fruits—whence its name. Honey is more than 40 percent fructose. Much sweeter than glucose and even than sucrose, it has a wide range of uses in the food industry. Indirectly, however, it is responsible for the disappearance of a part of the honey bee population in the United States, which raises the question in some people's minds whether glucose poses a danger to human health. Again, I must ask the reader's indulgence, for the opportunity to take up arms once more against the merchants of fear whom I assailed earlier is irresistible. The facts of the matter are as follows. Beekeepers, having learned that bees were attracted to fructose syrup that had been spilled in food-processing plants, naturally thought of using this inexpensive product to feed their hives. What they didn't realize (because no one realized it at the time) was that under sunlight fructose is gradually transformed into a compound called hydroxymethylfurfural, which, despite its delightful caramel odor, is poisonous to bees. But hydroxymethylfurfural does not pose a comparable risk to human beings, for the simple reason that humans are not bees. What's more, anyone who believes that hydroxymethylfurfural is to be avoided should also stop eating caramel—and indeed any dish in which liquids containing fructose have been heated or in which fruits and vegetables have been cooked. The idea is obviously absurd. This is why I feel an obligation, not only in my writings but also in my teaching and lecturing, to combat fears that have no basis in sound research and reasoned argument. The problem is not that we have too much chemistry, but that we do not have enough of it. Precisely because chemistry is a science, it permits us to distinguish

3.6 THE FRENCH CHEMIST EUGÈNE-MELCHIOR PÉLIGOT STUDIED MANY ORGANIC COMPOUNDS, INCLUDING CARAMEL.

between those things that pose dangers, when they are proved to exist, and those things that do not.

Let's move on now to sucrose—ordinary table sugar. This time it is clearly not necessary to make any introductions. I shall limit myself to saying that the sucrose molecule is a chain of glucose and fructose units, and that although sucrose does not play a role in Maillard reactions (about which I shall have more to say in a moment), it can be used to make caramel. This is a remarkable phenomenon, which can be extended to include other simple sugars. Heating table sugar to obtain caramel is not merely a splendid example of note-by-note cooking, but also an archetypal example because it shows that compounds do not need to be used only in their raw state; they can be cooked as well. In the case of caramel, especially if a drop of lemon juice or vinegar has been added, the sucrose molecules disassociate, splitting into glucose and fructose; the fructose molecules then react, forming fructose dianhydrides, reactive compounds that bond in turn with neighboring sugar molecules. By applying the same technique to glucose and fructose it is possible to obtain new kinds of caramel, which I have proposed naming "péligots" in honor of the French chemist Eugène Melchior Péligot, whom we have already met, a pioneer in the study of caramelization. More about these later.

There are a great many other simple sugars in addition to glucose and fructose, and it would be a wearisome business to go through them all, one by one. Let me limit myself therefore to mentioning just a few: mannose, rhamnose, galacturonic acid, guluronic acid, glucuronic acid, sorbitol, lactitol, maltitol, maltose, mannitol, xylitol, agarose, galactose. Some of them are today considered to be additives, others not, but the reasons why are of no concern to us here.

COMPLEX SUGARS

Simple sugars such as glucose, galactose, and fructose are technically known as monosaccharides. Sucrose, a disaccharide formed by linking two monosaccharides together, is somewhat more complex. Linking three monosaccharides together gives you trisaccharides, linking four together gives you tetrasaccharides, linking five together gives you pentasaccharides, and so on. Onions contain fructooligosaccharides, in which one glucose molecule is linked to several fructose molecules. Inulin, found in chicory, is also an oligosaccharide.

Once the number of monosaccharide units exceeds ten, we arrive at the complex sugars, polysaccharides. Although polysaccharides generally are tasteless, they are useful in creating consistency, as we saw not only in the case of cellulose, but also of amylose and amylopectin, found in starches; of chitosan, produced from chitin; of various compounds extracted from algae; and of gums, to name only a few. As I say, these compounds typically have no taste except when small amounts of sugar are released in the mouth. Salivary amylases, for example, are proteins that function as enzymes, cutting up long chains of amylose and amylopectin in flour, with the result that after a few

3.7 MONOSACCHARIDES (REPRESENTED HERE BY POLYGONS) CAN BE JOINED TOGETHER TO FORM DISACCHARIDES, TRISACCHARIDES, TETRASACCHARIDES, AND POLYSACCHARIDES.

seconds a sweet taste appears—enough of one, at any rate, to allow an inventive cook to achieve novel effects.

ALCOHOLS AND POLYOLS

From sugars (technically, polyhydroxy aldehydes) we now go on to alcohols and polyalcohols (or polyols). These compounds are important not only for the taste of foods, but also for their consistency.

ALCOHOLS

The best-known alcohol is ethanol, preponderant in wines, beers, ciders, and brandies. It was the first alcohol to be distilled in the early medieval period—a discovery that bought about an upheaval in chemistry, for until then it was believed that all matter was made of four primary substances (water, fire, earth, and air). Distillation had succeeded in creating a fifth substance, a *quinte essence.*

Ethanol is nevertheless not the simplest alcohol from the point of view of its molecular structure. In organic chemistry, an alcohol defined is a compound whose molecules bear a hydroxyl group (in which one oxygen atom is bound to a hydrogen atom). The simplest alcohol—having a single carbon atom, a functional group (in this case the hydroxyl group $-OH$), and two hydrogen atoms for good measure (satisfying the requirement that each carbon atom must be bound to four other atoms)—is methanol, also known as wood alcohol. It is prepared by prolyzing wood, which is to say that the wood is heated in an airtight stove, producing gases that are highly flammable because they are rich in methanol. I should hasten to add that methanol is very toxic (indeed, in sufficient quantities it makes people go mad and blind), and government agencies are therefore justified in strictly regulating the production and sale of white alcohols, which give off methanol during the distillation process. The small quantity of methanol that survives this process is at least partly responsible for hangovers. Drinking too much white alcohol causes the jaws to tighten and become clenched—the first sign of methanol poisoning. This may well be the origin of the French expression for a hangover, *gueule de bois* (literally "wooden mouth"), though of course no one can say for sure.

Adding a second carbon atom gives us ethanol, a good example of a compound that has both a taste and an odor. Ethanol is a wondrous thing for purely intellectual reasons as well: neither sweet nor salty nor sour nor bitter, it once again refutes the false theory of the four tastes; more prosaically, perhaps, its physiological effects are somehow connected with the fact that ethanol seems to act on the trigeminal nerve, a nerve that comes from the rear of the skull and whose endings in the nose and the mouth detect pungency and

TASTE · **139**

coolness. And because ethanol is a colorless liquid that is less dense than oil, it has the very practical virtue of allowing us to create layered cocktails, beginning with the famous Irish coffee. Human beings value ethanol especially for its psychoactive effects, of course, but they are not alone: even horses and cows are fond of fermented fruits!

Whereas methanol molecules are built around a single carbon atom, ethanol molecules contain two of them. As the number of carbon atoms gets larger, the variety of alcohols continues to grow: they come in every size, with names such as "propanol," "butanol," "pentanol," "hexanol," "heptanol," and "octanol." The list goes on, for the chain of carbon atoms that forms the backbone of these molecules can be almost endlessly elaborated with branchings and various other embellishments. Earlier I mentioned 1-octen-3-ol (octenol), which has a marvelous odor of undergrowth—hence, its colloquial name, "mushroom alcohol." The root *oct-* means that there are eight linked carbon atoms; the suffix *-en* means that two carbon atoms (whose position is indicated by the "3") are joined by two bonds instead of one; *ol* designates the hydroxyl group typical of alcohols; the 1 specifies a further modification of the molecule's structure at the first position. Once again, this alcohol is not merely sapid, but fragrant as well.

The world of alcohols is immense, and we risk getting, well, a headache if we go on cataloging them much longer. Let's move on, then, to polyols, at least a few of which are already known to some modern cooks.

POLYOLS

Six polyols are commonly used by the food industry: sorbitol, mannitol, maltitol, isomalt, xylitol, and lactitol. Sorbitol, mannitol, and xylitol occur naturally in various fruits and vegetables, but samples that are obtained artificially through the hydrogenation of sugars come from starches. Thus sorbitol is derived from glucose, manitol from fructose, and maltitol from maltose.

Polyols can be used as sweeteners, but also as emulsifiers, stabilizers, moisturizers, thickeners, and texturizers. Unlike sucrose they do not cause dental cavities, and their calorie content is low—whence their appeal as an ally in combating obesity.

Isomalt (E953) is a combination of two hydrogenated saccharides: alpha-D-glucopyranosyl-1,6-sorbitol and alpha-D-glucopyranosyl-1,6-mannitol. It is obtained from sucrose in two steps: first, with the help of enzymes, the sucrose is transformed into isomaltulose, which then is hydrogenated to form isomalt.

Lactitol (E966) is a hydrogenated disaccharide obtained by lactose hydrogenation.

Maltitol (E965) is obtained by catalytic hydrogenation of D-maltose, a disaccharide present in malt that is made artificially through the saccharification of purified starch using an enzyme known as beta-amylase.

Mannitol (E421) owes its name to manna, the sweet-tasting exudate of the ash tree (*Fraxinus ornus*), of which it is the principal constituent. It is also found in olives, figs, and the sap of the larch tree, as well as in some edible mushrooms and in seaweeds of the genus *Laminaria*. Catalytic hydrogenation transforms a mixture of D-glucose and D-mannose into a mixture of D-mannitol and D-sorbitol.

Similarly, D-sorbitol takes its name from the berry of the rowan or mountain ash tree (*Sorbus aucuparia*), from which it was first extracted in 1872. A sugar found in many vegetables, it too is produced commercially through catalytic conversion of D-glucose (dextrose), itself obtained through enzymatic hydrolysis of starch.

Last but not least, xylitol (E967) is found in many fruits (greengages, strawberries, raspberries, and the like) as well as in cauliflower. It is commercially produced through catalytic hydrogenation of D-xylose, which is abundant in corn cobs, almond hulls, and birch bark. Xylitol, like sorbitol, has a refreshing effect on the palate.

INTENSE SWEETENERS

Sweetening compounds are legion, but the key word here is the adjective *intense*. Some compounds are several thousand times sweeter than sucrose. The degree of sweetness is crucial, for intensely sweet substances sweeten while providing little or no energy, which is to say they contain few, if any, calories needing to be burned off—a considerable advantage at a time when the plague of obesity shows no sign of abating.

No one knows to what extent traditional Western diets, which developed in an age when famines were still common, are responsible for our present predicament. It is plain, however, that unless food suddenly ceases to be overabundant, eating habits will have to change if we are to avoid dying from cardiovascular disease in even greater numbers than we do today. From this point of view, note-by-note cooking has undoubted advantages. There is no reason why we should not make use of intense sweeteners, for innumerable studies have established that the toxicological risk associated with them is too small to warrant limiting commercial production, much less banning it altogether. All the compounds I discuss here are currently approved for commercial use in both Europe and the United States.

Acesulfame potassium, also known as acesulfame K (E950), was discovered by accident in 1967 by two agricultural research chemists at Hoechst AG in West Germany, Karl Clauss and Harald Jensen, and described in a paper they published six years later. Clauss happened to dip his finger into a batch of chemicals without noticing and then, licking it to pick up a piece of paper, tasted a mixture of remarkably sweet compounds. Together with Jensen, he managed to isolate the sweetest of them (two hundred times sweeter than sucrose), which also turned out to be the easiest one to use, acesulfame K. Molecularly very stable, it withstands both pasteurization and sterilization, decomposing only at temperatures higher than 200°C (392°F). The initial sensation of sweetness is followed, however, by a bitter aftertaste.

Aspartame (E951) was also discovered by accident two years earlier, in 1965, by James Schlatter, a chemist at G. D. Searle & Company, who was studying oligopeptide synthesis as part of an attempt to develop antiulcer drugs. Moistening his finger in order to turn the pages of a journal, he realized that he had unwittingly touched a sweet substance. Although aspartame is one of the most thoroughly tested food additives on either side of the Atlantic, it still comes under attack from time to time. The compound's charms are nevertheless undeniable: a sweetening power 130 to 200 times greater than that of sucrose and a similar taste profile. It is unstable when exposed to prolonged heating, which releases phenylalanine and methanol, but the small quantities in which it is typically used in cooking make it no more dangerous than fruit jams, whose pectins are likewise sources of methanol.

Cyclamic acid and its sodium and calcium salts (cyclamates, for short), bear the number E952. Like so many intense sweeteners, cyclamates were also discovered accidentally, in 1937. Michael Sveda, a graduate student at the University of Illinois who was engaged in research on the synthesis of fever-reducing drugs, discovered that a cigarette he was smoking had a sweet taste after he picked it up from the lab bench.

Neohesperidine dihydrochalcone (E959) was discovered in 1963 by two American researchers, Robert Horowitz and Bruno Gentili, who were study-ing the relationship between the molecular structure of citrus compounds and bitter taste. Horowitz and Gentili had identified neohesperidine as one of the flavonoids found in bitter (Seville) oranges. To their astonishment, a derivative of this compound, neohesperidine dihydrochalcone, turned out to be sweet rather than bitter. It has an unpleasantly pronounced and persistent cool, licorice-like aftertaste, however, which can be offset only in combina-tion with other sweeteners. Neohesperidine dihydrochalcone is also used for a purpose that falls outside the scope of the present chapter: in concentra-tions lower than those that make the sweet taste appear, it enhances odors.

Saccharin and its sodium, potassium, and calcium salts are assigned the number E954. Saccharin, the first of the intense sweeteners to be discovered, was accidentally encountered in 1878 by two chemists at Johns Hopkins, Ira Remsen and Constantin Fahlberg, who were investigating a class of coal tar derivatives known as toluene. At supper one evening after a day in the lab, Fahlberg happened to notice that his fingers tasted sweet. Commercial pro-duction by Monsanto Company began a quarter-century later, in 1902. Sac-charin has a metallic and bitter aftertaste, less pronounced in the calcium salt than the others.

Sucralose (E955) was discovered almost a hundred years after saccha-rin, in 1976, in London. This time the discovery was the result of a misun-derstanding: a chemist tasted a group of sugar derivatives instead of *testing* them, as he had been instructed to do. On further investigation he found that these halogenated sucrose derivatives (in which organic compounds form bonds with iodine, chlorine, and fluorine) include an insipid series, a bitter series, and a sweet series. Trichlorogalactosucrose, a chlorinated derivative of sucrose valued for its intensely sweet flavor, is now conventionally known as "sucralose."

Thaumatin (E957) is an outlier on this list: it is a protein and therefore a large molecule. Thaumatin is extracted from the katemfe fruit (*Thaumatococcus danielli*), found in West Africa. The fruit itself was already known for its intensely sweet flavor, but it was only in 1972 that two sweet proteins, thaumatin I and II, were isolated by a pair of Dutch chemists, Henrik Van der Wel and Kees Loeve. Alas, it has a very strong licorice aftertaste, like rebaudioside A, extracted from the stevia plant, a native of South America, which quite recently has been recognized as an intense sweetener.

The sale of a number of intense sweeteners is prohibited in Europe, such as neotame (made by the American company NutraSweet) and alitame (made by Pfizer).

FLAVORING AGENTS

Earlier I explained why it is essential to distinguish between flavor (*goût* in French), a synthetic sensation, and taste (*saveur*), which is my subject in this chapter. Neither commercial producers nor regulatory bodies have yet really grasped the distinction, with the result that incoherence continues to reign, even in the scientific literature. I therefore persist in taking a clear, consistent, logical line in this matter, in the possibly forlorn hope that one day everyone will see things my way.

The category of additives coded E600–699 bears the official name "Flavor Enhancers." Surely there can be no objection to speaking of taste enhancers or odor enhancers. But flavor enhancers? One commonly hears it said that salt, for example, is a flavor enhancer. It is true that an unsalted vegetable soup does not have a vegetable flavor; it only acquires one when salt is added in a quantity that nonetheless does not make the soup taste salty. Salt is therefore said to bring out the flavor of vegetables. In that case, is it the odor that is brought out or the taste? No one knows—although there is reason to believe that it may be odor because odorant compounds, which do not dissolve in water, dissolve even less easily when the water is salted—and so are able to be "salted out," or extracted, from the liquid into the air.

However this may be, salt has another remarkable effect that can be experienced by conducting a test with tonic water, which contains quinine: add a bit of salt, and the water becomes sweeter and less bitter. In other words, salt is a taste enhancer because it enhances the sweet taste of the sucrose; but it is also a taste weakener, because it weakens the bitter taste of the quinine (which perhaps is why some people lightly salt their coffee). At all events, salt cannot be said to enhance a flavor if it diminishes a taste.

Even if the category of artificial flavoring substances that we are presently considering has been misnamed, it includes compounds that the note-by-note cook will want to make use of. I have already mentioned monosodium glutamate. It is the sodium salt of glutamic acid, an amino acid that is commercially produced by fermenting molasses and other liquids obtained from starch sugars. This acid, in the L form, is very commonly found in the natural world. It is salted out by the sodium ion, then purified and crystallized in the form of odorless white crystals. Depending on the individual, the sodium salt has a sweet or salty taste (some detect the taste of chicken broth); in every case, it intensifies the sensation of a particular taste. Other substances are reputed to have similar effects, such as sodium guanylate, sodium inosinate, isovaline, sodium DL-threo-8-hydroxyglutamate, sodium DL-homocysteine, sodium L-aspartate, sodium L-alpha-aminoadipate, L-tricholomic acid, and L-ibotenic acid. All these substances are derived from amino acids, and their action remains poorly understood.

Maltol, ethyl maltol, and furaneol enhance the fruit taste of jam and caramel and the rich, round notes that intensify the sensation of sweetness. Various substances affect the sensation of milk flavor: dioctyl sodium sulfosuccinate modifies the fresh milk taste, and NN-diorthotolylethylene diamine and cyclamic acid modify the butter taste. These compounds act in infinitesimally small doses, on the order of 0.0001 parts per million. Finally, let me mention diethyl glutamate, which can augment sweet and bitter tastes as well as the smell of ether, and methylpropyl alcohol, which acts in a way similar to sodium glutamate.

BITTERANTS

Is bitterness a sign of danger? It may be. Earlier I mentioned that many bitter alkaloids (compounds, extracted from plants, whose molecule contains one nitrogen atom) are toxic. Nevertheless we are not just animals—and even our animality is singular. One has only to look at the many different sorts of food people eat to see that bitterness is by no means avoided. Beers are bitter. So are browned onions, to say nothing of the many aromatic plants, such as rue, that have been used in cooking since medieval times.

Nearer to our own day, the chef Édouard Nignon (1865–1934) sang the praises of bitter tastes in his remarkable book *Éloges de la cuisine française*, published a year before he died. Note that I speak of bitterness in the plural: like every good cook, Nignon knew (in his case without having to wait for the confirmation provided by biological research in the past few decades) that there are various kinds of bitterness, some persistent, others not; and that they stimulate different parts of the mouth and excite very specific sapictive receptors on the papillae.

In the food industry, the most commonly used bitter compounds are quinine (and its salts) for sodas and alpha-humulene for beers, but the rinds of citrus fruits contain many interesting compounds, among them naringin, which is found in grapefruit.

EXTRACTING BITTER COMPOUNDS

One of the difficulties facing note-by-note cooks today is that pure compounds are sometimes hard to come by. The Internet makes it possible to buy products from vendors halfway around the world, but it is worth keeping in mind that, if need be, we can always extract compounds ourselves. To take only the simplest example, heating grapefruit rinds in water produces a very bitter solution. Typically what is extracted in this way is a mixture of compounds, of course, rather than a pure compound. The technique of liquid–liquid extraction is commonly used in laboratories to separate the component parts: one liquid containing several dissolved compounds is combined with another liquid that does not mix with it; after the two solvents have been vigorously shaken, the compounds that dissolve more readily in the second one migrate to it.

In the kitchen, all you need are a glass jar and some water and oil. Take a grapefruit, for example, grind it up, and then put the crushed rind and flesh in the jar with equal amounts of water and oil. Screw the top on and shake the jar vigorously. Let the solids settle, and then separate the oil from the water. Taste the two liquids: each will have its own distinctive flavor! I leave it to you to try the same experiment with tomatoes (raw or cooked), fruits, or meats (again, raw or cooked). Whatever ingredients you choose, from one flavor you will obtain two new flavors. The water and oil that have been flavored in this way do not contain just one compound; they are what chemists call "fractions," comprising all the compounds that these liquids dissolve respectively. There is no reason why we should not make use of fractions as well.

MATRIX EFFECTS

By now we have learned a few of the notes—sapid notes—that can be played on this marvelous piano we call cooking. It remains for us to mix them together. How do we go about doing this?

It would be altogether presumptuous for a chemist to give instructions in this matter, for how best to arrange the notes is an artistic, not a technical question. Even so, the chemist should not be bashful about calling attention to facts that may be useful to culinary artists. In particular, chefs should be mindful that the taste of a dish is not reducible to an isolated or momentary sensation. A dish must be constructed in such a way that its sensory effects will be registered over a succession of moments since the perception of flavor is an enduring sensation—or should be one if it is not.

With regard to the water in which sapid compounds are dissolved, the taste one perceives lasts only as long as the longest lasting of the sensations produced by the compounds that are mixed together in the water. But just as note-by-note cooking cannot be reduced to liquids alone, so there is no reason why the taste of a note-by-note dish has to be brief. The idea of producing

solutions and nothing else holds no interest in any case. With enough imagination and ingenuity, the overall sensation can be very intricately structured,
particularly if we exploit the properties of colloidal systems such as the ones
we examined in chapter 2.

The key to success in this endeavor is to be found in what I call "matrix
effects." Let's once again consider sodium chloride—the salt we use every
day in our cooking. Depending on whether we are dealing with coarse salt
or finely granulated salt or rock salt or hand-harvested sea salt (*fleur de sel*)
or flake salt or the golden pyramid-shaped crystals known as Cyprus salt, the
taste is not always the same, nor does it contribute to the perception of flavor
in the same way because the chloride and sodium ions are not released at the
same rate for every shape.

Varying the shape and size of different sorts of particulate matter is not
the only possible way to shorten or lengthen the perception of taste and flavor, however. Dissolve some salt in water and now give the solution the form
of a gel—by adding gelatin, for example. The particular microstructure, or
"matrix," of the gel causes the salt to be released into the gelatinized water
more slowly; in other words, it produces a different salty sensation. We can
introduce an additional degree of complexity by making the gels more or
less firm, using gelatin as before or some other gelling agent. In this way a
whole range of release rates for the salt can be devised. If they are skillfully
combined it will be possible to make dishes having an immediate sensation of
saltiness that gradually disappears and then suddenly returns.

These mechanisms are simple enough. But matters become rather more
complicated once one is acquainted with chemical bonds—particularly, in the
case of salt, once one knows how to combine ions (in this case chloride and
sodium) with various molecules, whether these molecules are added in solution to the liquid trapped in the matrix of a gel or whether they themselves
constitute the solid walls of the gel's matrix.

All this goes to show that note-by-note cooking can attain an unrivaled
degree of precision with respect to taste. Something similar may in fact be
possible using the methods of traditional cooking, but only with immensely
greater difficulty and without being able to separate tastes from odors. With

note-by-note cooking, by contrast, the independence of sensory registers—think of them by analogy with the registers of an organ, a set (or "rank") of pipes—multiplies occasions for creativity by whole orders of magnitude!

A NEW BASIC TASTE

The world of tastes is still, as I say, poorly understood. Should we deplore this state of affairs, or should we rejoice in the opportunities it provides for making new discoveries? Optimists will prefer to adopt the latter view, and their hopes will grow further when they learn that fats, which have long been reputed to be flavorless, turn out to be rather less insipid than was believed only twenty years ago.

More precisely, what has been discovered is a new modality or sensory faculty: human powers of gustatory perception include the ability to perceive the taste of what are technically known as long-chain unsaturated fatty acids. What exactly is meant by this term? Let's begin with an oil. We saw earlier that its constituent molecules, triglycerides, resemble microscopic combs with three teeth. Such molecules can be formed by causing a reaction between glycerol (the shaft of the comb) and fatty acids (the teeth). Fatty acids? Think of a linear chain, a sort of backbone, formed by joining carbon atoms together. To one of the last carbon atoms let's attach a carboxyl ($-COOH$) group, in which a carbon atom is bound to an oxygen atom on the one side and to a hydroxyl group (an oxygen atom and a hydrogen atom) on the other. Next, let's attach hydrogen atoms to the carbon atoms so that each of the carbon atoms is duly bonded to four other atoms in all. We now have a saturated fatty acid. To get an unsaturated fatty acid, remove one hydrogen atom from two neighboring carbon atoms and join these carbon atoms by means of a new bond.

Chemists in France recently discovered that such compounds are "recognized" by sapid receptors on the papillae. Initial studies were conducted on mice, but experimental evidence regarding human perception continues to accumulate. At first it was unclear whether one was dealing with a taste—the taste of fat—or with something else. It is now quite clear, however, that fatty matter does not consist only of a group of compounds that we appreciate by virtue of a sort of nutritive reflex, so that eating fat triggers an agreeable sensation like the one provoked by the ingestion of modified starches, which slowly release glucose into the digestive system. No, the perception of fat does seem in fact to constitute a new sensory modality—a sixth basic taste, as it is often referred to.

In addition to the compounds I have discussed in this chapter, there are many others, natural and synthetic alike, that are now known to have novel and distinctive tastes. How many other such discoveries await us in the years ahead?

FURTHER
READING

Breslin, P. A. S., and G. K. Beauchamp. "Salt Enhances Flavour by Suppressing Bitterness." *Nature* 387 (1997): 563.

Chandrashekar, J., D. Yarbolinsky, L. von Buchholtz, Y. Oka, W. Sly, N. J. P. Ryba, and C. S. Zuker. "The Taste of Carbonation." *Science* 326 (October 16, 2009): 443–445.

Chaudhari, N., A. M. Landin, and S. D. Roper. "A Metabotropic Glutamate Receptor Variant Functions as a Taste Receptor." *Nature Neuroscience* 3, no. 2 (2000): 113–119.

Doty, R. L., ed. *Handbook of Olfaction and Gustation.* New York: Marcel Dekker, 1995.

Faurion, A. "Naissance et obsolescence du concept de quatre qualités en gustation." *Journal d'agriculture tropicale et de botanique appliquée* 35 (1988): 1–19.

Hladik, C. M., and B. Simmen. "Taste Perception and Feeding Behavior in Non-human Primates and Human Populations." *Evolutionary Anthropology* 5 (1996): 161–174.

Laugerette, F., P. Passily-Degrace, B. Patris, I. Niot, J.-P. Montmayeur, and P. Besnard. "CD 336, un sérieux jalon sur la piste du goût du gras." *M/S: Médecine sciences* 22, no. 4 (2006): 357–359.

Nelson, G., J. Chandrashekar, M. A. Hoon, L. Feng, G. Zhao, N. J. Ryba, and C. S. Zuker. "An Amino-acid Taste Receptor." *Nature* 416 (2002): 199–202.

Rolls, E. T. "Mechanisms for Sensing Fat in Food in the Mouth." *Journal of Food Science* 77, no. 3 (2012): S140–S142.

Tucker, R. M., and R. D. Mattes. "Are Free Fatty Acids Effective Taste Stimuli in Humans?" *Journal of Food Science* 77, no. 3 (2012): S148–S151.

Uziel, A., J. G. Smadja, and A. Faurion. "Physiologie du goût." In *Encyclopédie médico-chirurgicale*, Otorhino-laryngologie, 2, 20490 C10. Paris: Éditions techniques, 1987.

FOUR ODOR

NO ONE WILL DISAGREE that aromatic plants have an aroma, that wines have a bouquet, that flowers have a fragrance—and that meats have an odor. In reality, however, meats, like all foods, have two odors: an "orthonasal" odor that is perceived by smelling and a "retronasal" odor that appears when the odorant compounds released by chewing pass through the air inside the buccal cavity and then rise up into the nose through the retronasal passages.

But why aren't retronasal and orthonasal odors the same? After all, the odorant compounds are the same in each case. There are two other things to consider, however. First, that certain compounds are released only when a food is destructured. Second, that the perceptible proportion of each odorant compound varies depending on whether it is smelled or chewed because chewing heats the food to a temperature of 37°C (98.6°F) or so, which modifies the released compound's odor profile. This profile is what we recognize when we eat.

An odor can be thought of as resembling the shape of a tent. The way in which poles of different heights are placed to support the tent fabric is what gives the tent its shape. If you change the number or the position or the height of the poles, you get another shape—another odor. Thus if you add a little orange blossom water (one group of poles) to a strawberry (another group of poles), you get a new smell, the smell of wild strawberries. So, too, if you add a drop of pastis to a small amount of coffee, an odor of licorice appears. Sensory physiology provides a wealth of evidence to suggest that this phenomenon is responsible for the discovery of many classic combinations in cooking: chestnut and fennel, carrot and orange, and so on.

MANIPULATING ODORANT COMPOUNDS

Nothing is more human than the urge to combine different fragrances and, more generally, the passion for modifying the basic flavors of foods. Since antiquity, cooks and perfumers have experimented with herbs, spices, flowers and other parts of plants, and various animal substances. It is sometimes said that our ancestors were drawn to these products because they have antiseptic properties analogous to those of modern preservatives. It is true, for example, that rosemary contains rosmarinic acid, which is a good preservative; also that thyme contains thymol, that clove contains eugenol, that hot peppers contain extremely pungent capsaicinoids that kill microscopic mushrooms, that many vegetable terpenes have antibacterial properties, and so on. Nevertheless there are good grounds for supposing that humans were attracted in the first place by the odor of these ingredients. Exactly why is still unknown, however. Why do we like the smell of roses? This remains a mystery, but molecular biologists are now beginning to shed light on the question through the study of pheromones, which trigger specific behaviors in animals, and of the role odors play in the immune system.

All this is a very complicated story that has little immediate relevance to cooking. We need only assume for present purposes that our earliest ancestors didn't have to be terribly astute to notice that foods take on the odor of vegetable and animal tissues that are added to them. It was only much later, probably in the Neolithic period, that techniques were invented for making nutritive oils, some of which have very distinctive flavors. One thinks in particular of oils made from hazelnuts, walnuts, pistachios, and olives. Maceration, infusion, and other types of decoction probably came after. And yet here again it would have readily been appreciated that the odor of fragrant substances is released in both water and oil. Enfleurage, a technique that is still used today to make perfumes from delicate flowers, involves solid fats, but the basic idea is the same. I shall come back to it in due course.

Distillation is more complex. It seems to have first appeared with the Chaldeans around the sixth millennium B.C.E. Once more we encounter an invention that demanded not so much genius as well-developed powers of observation. It would not have escaped anyone's notice that heating fragrant substances gives off fragrant fumes, for example, or that placing a cold solid above boiling water causes the vapor to condense into water, which can be collected from the surface of the solid by tilting it. A closed vessel designed for this latter purpose would have been an early prototype of the modern distillation device known as a retort. The procedure seems to have been improved in the ninth century by the Arabs, to whom we owe the word *alembic*, but not much more progress was made until the thirteenth century, when pharmacists began to prepare medicinal oils and ointments. Many essential oils used still today by perfumers and aromaticians (who combine such oils with herbs, spices, and various synthetic compounds to improve the taste and smell of foods) were first distilled by pharmacists in the sixteenth and seventeenth centuries. In the first half of the nineteenth century, artisanal manufacture of such products gave way to industrial production.

Improved distillation columns made it possible to separate mixtures of odorant compounds. We need not dwell on the details here, but even as brief a survey of odorant compounds as this one cannot omit to mention the discovery by chemists in the nineteenth century that simple organic compounds have agreeable odors, whether they are synthesized or extracted from natural substances. In 1834, Eugène-Melchior Péligot and another French chemist, Jean-Baptiste Dumas (1800–1884), isolated cinnamaldehyde from cinnamon oil. (If you use an atomizing sprayer to lightly coat toasted flour with cinnamaldehyde, professional cooks will tell you that they smell cinnamon.) Three years later, in 1837, the German chemists Justus von Liebig and Friedrich Wöhler isolated benzaldehyde from the oil of bitter almonds. The first aromatic oils were synthesized in the 1840s from short-chain fatty-acid esters and various alcohols that had attracted attention because of their fruity odor. Methyl salicylate (artificial wintergreen oil) was introduced in 1859, and benzaldehyde was synthesized in 1870. With the industrial synthesis of

VANILLIN

4.1 A VANILLIN MOLECULE. THE COMPOUND HAS A LOVELY AROMA, SIMILAR TO THAT OF VANILLA. THE TWO ARE NEVERTHELESS NOT TO BE CONFUSED SINCE VANILLA CONTAINS MANY OTHER ODORANT COMPOUNDS THAT ENRICH THE AROMA OF THE FERMENTED PODS OF THIS CLIMBING ORCHID.

vanillin in 1874 by Haarmann & Reimer GmbH, and then of coumarin in 1878, a new sector of commercial production opened up. New developments have been regularly forthcoming ever since with regard not only to the isolation of natural compounds, but also to the synthesis of compounds known to exist in nature as well as previously unknown compounds.

METHODS OF EXTRACTION AND PROCESSING

I have already mentioned a few of the techniques used by chemists and perfumers to produce odorant compounds, whether by extraction from natural substances or by synthetic preparation. Over the centuries these methods have been refined to one degree or another. The most recent advance is due to the use of carbon dioxide in a so-called supercritical state—that is, a state of matter that may assume the form of either a gas (a highly fluid substance that is introduced into the animal or vegetable tissues from which one wishes to extract odorant compounds) or a liquid (in which such compounds are dissolved). Each extract has its own distinctive characteristics. For the moment, however, let's focus our attention on more traditional methods of extraction and then quickly take a first look at processed products and synthesized compounds.

EXTRACTIONS

Ointments and pomades are examples of a fatty preparation containing odorant compounds that has been produced by cold or hot enfleurage. In hot enfleurage, fragrant substances are directly immersed in a hot liquid obtained by heating solid fats; in cold enfleurage, the procedure is more involved since it is necessary first to cover a neutral fat with a fragrant substance, then wait until the odorant molecules pass into the air and from the air into the fat. Industrial production in each case was inaugurated in the south of France in the nineteenth century.

Essential oils, or essences, may likewise be obtained by either cold or hot methods. The former involve a series of mechanical procedures (the washing and pressing of citrus peels, for example); the latter involve a process known as steam distillation, in which steam is injected into an aromatic substance to trigger the release of odorant compounds; then, once the vapor has condensed, the oil is separated from the aqueous phase. Using steam distillation, the yield is somewhere between 0.1 and 1 percent. Although the essential oils extracted in this way are concentrates of aromatic substances, they need not have all the properties of cooking oil. A drop on a piece of paper, for example, does not always leave a stain.

Essential oils are to be distinguished from distillates, which contain ethanol (ethyl alcohol) and are obtained by distilling vegetable matter with ethanol or with a hydroalcoholic solution. The principle is the same in either case: evaporating a solvent produces odorant compounds, which are collected by allowing the vapor to condense.

Concrete essences (concretes), such as oleoresins, are prepared by extraction using nonpolar solvents (toluene, hexane, petroleum ether). The odorant compounds are subsequently recovered with the elimination of the solvent by means of filtering and distillation.

Absolute essences (absolutes) are obtained by heating and stirring concretes with an alcohol, typically ethanol. Fractional distillation of this mixture yields a wax-free absolute once the ethanol has been eliminated.

Resinoids are the result of treating the exudates of vegetable tissues with solvents such as methanol, ethanol, and toluene. They are not to be confused with oleoresins, which are concentrates prepared by solvent extraction, often from spices.

Tinctures, or infusions, are solutions in which ethanol or a hydroalcoholic solution is used as a solvent to produce liquid concentrates (typically of fruits) with the aid of various filtration techniques, sometimes accompanied by the recovery of very volatile compounds.

Alcoholic distillates, infusions, macerations, percolates, and the like have been used since ancient times. Pure substances (menthol, for example) are sometimes obtained by crystallization.

PROCESSED PRODUCTS

If the food industry produces fragrances and flavoring agents through the extraction of naturally occurring odorant compounds, it also creates them artificially by processing vegetable and animal matter in various ways. The phenomenon of Maillard reactions, for example, can be exploited to impart the smells of grilled and other kinds of cooked foods. Enzymes can be used to hydrolyze various substances. Fermentation yields a vast range of odors. Thus, for example, the equivalent of vanillin can be obtained by fermenting pine needles with the aid of microorganisms that are likewise found in nature.

SYNTHESIS

Finally, alongside compounds extracted from natural substances or created on the basis of such substances, there are synthesized compounds. Once again, let us be clear about what is meant by the term *synthesized*: synthesized (or synthetic) compounds may be identical to natural compounds or they may be altogether new compounds.

The manufacture of perfumes, which many people consider a frivolous indulgence, nothing more, has nevertheless become a very sizeable and prosperous industry that attracts some of the best young minds in organic chemistry. Several Nobel prizes have already been awarded in recognition of advances in this domain, and attempts to create new fragrances have led to the discovery of mechanisms of molecular rearrangement ("chemical reactions," in the more familiar phrase) having a variety of applications in fields beyond perfumery. More than ten thousand odorant compounds have been identified so far—and we have not yet begun to approach the limits of molecular synthesis!

NATURAL, SAME AS NATURAL, ARTIFICIAL

After mentioning benzaldehyde and vanillin, I broke off my historical précis of naturally occurring odorant compounds that can also be synthesized. The procedure for making such substances in the laboratory is so simple that they are within the reach of any trained herbal chemist. And yet despite the current vogue for "natural" products, the ability to synthesize compounds that are in every respect identical to compounds extracted from plants or otherwise present in the environment poses problems, at least from the regulatory point of view.

Once again, we need to be clear at the outset about the meaning of the terms we use. Things are properly said to be "natural" if they have not been transformed by human activity; things that have been transformed by human activity are said to be "artificial." This means that cultivated plants (carrots, onions, and so on) are not natural, nor are the compounds extracted from plants. Myristicin, for example, which gives nutmeg a significant part of its flavor, is natural only so long as it resides in the kernel of the nutmeg tree's fruit, not if it has been extracted from a nutmeg pod. Nor is an essential oil prepared by human hands natural, even if a government regulatory agency says it is. The molecules that constitute synthesized myristicin, by contrast, are exactly the same as those that constitute the myristicin found in a nutmeg pod. This is what is meant when synthetic myristicin is said to be indistinguishable from natural myristicin—"same as natural" in the conventional shorthand.

And yet there is nothing wrong with the term *synthetic*. Indeed, in debating such matters and in drafting regulations, we would do well to use it, for it has the virtues of both accuracy and transparency. No matter that myristicin is still myristicin whether it is synthesized or extracted, citizens in a democratic society have a right to be fully informed about the nature and provenance of the products they buy. Benzaldehyde, if it is obtained by using specially selected reagents to cause a chemical reaction under specific laboratory conditions, is a synthetic compound. It is therefore artificial as well, and there is no reason why the manner in which it was produced should not be plainly stated.

So long as these terms are used ambiguously, however, it is understandable that people should regard the "aromas" created by the "aromatics" industry (misnomers both, for the reasons I gave earlier) with suspicion. All such products are artificial, in the strict sense of the word, because they have been prepared by human beings. Some of them contain compounds extracted from plants and animals; others contain synthesized compounds. Public trust in the wholesomeness of processed food products would be greater, I believe, if we were to speak instead of "fragrant preparations"—or, better still, of "odorigenic compositions," as I shall explain in due course.

VOLATILITY, THRESHOLD PERCEPTION, TOXIC RISK

Just as sapid compounds must be soluble in water, since they have to pass through saliva in the mouth in order to reach the papillary receptors, so too odorant compounds must be able to be released into the air. This is what is called "volatility," which depends both on the particular chemical characteristics of the atoms present in the molecules and on the size of these molecules.

It used to be believed that olfaction is a rather straightforward business: once released, odorant molecules bind directly with receptors, one by one. But the simple picture of an odorant molecule as a key that fits into a receptor lock, which we looked at in chapter 3, has to be modified in light of the recent discovery of proteins in nasal mucus that serve as intermediaries between odorant molecules and receptors. Current research may be expected to further revise our understanding of this phenomenon. Whatever the exact mechanism of olfaction turns out to be, however, it will be useful for note-by-note cooks to be familiar with the concept of the threshold perception of an odorant compound. This is the smallest concentration in which the odor of a compound can be perceived under a specified set of experimental conditions. For certain compounds (1-p-menthene-8-thiol, for example, which is found

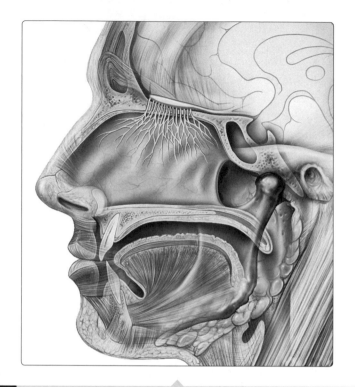

4.2 VOLATILE COMPOUNDS REACH OLFACTORY RECEPTORS IN THE NOSE THROUGH THE NOSTRILS (ORTHONASAL ROUTE) OR THROUGH THE RETRONASAL PASSAGES ABOVE AND BEHIND THE MOUTH. WHEN A COMPOUND BINDS WITH A RECEPTOR, WE ARE ABLE TO DETECT IT AS AN AROMA. (THE WORD AROMA IS PROPERLY USED TO REFER TO A FRAGRANCE EMITTED BY AN AROMATIC PLANT, BUT THIS RULE IS FREQUENTLY IGNORED TODAY.)

in grapefruit), the threshold is very low—so low, in fact, that a teaspoonful in enough water to fill a million swimming pools would still be detectable!

The power of odorant compounds acts also as a deterrent, preventing us from ingesting them in excessive quantities. We have seen that some compounds are truly dangerous. Estragole, for instance, found is tarragon and basil, is converted by the human body into hydroxyestragole, a carcinogenic and teratogenic substance even in very small doses. The fact that odorant compounds are typically found in very small concentrations in vegetable matter reduces their noxious effects, though the synthetic preparation of these compounds in concentrated form is evidently not without its own dangers. Cooks must therefore choose between learning the art of distillation or turning to professional distillers who can do the job for them. Many qualified experts recommend that odorant-compound concentrations not exceed ten to twenty parts per million. What does that mean as a practical matter? It means that a teaspoonful of an odorant compound—a few grams—will need to be dissolved in a million grams (roughly a ton) of an inert solvent to yield a tolerably safe dose. Not all inert solvents are suitable, by the way. Water, for example, isn't a good choice for this purpose. Cooking oils are frequently used, but a variety of alcohols, including ethanol and propylene glycol, may also be considered.

A word of advice for cooks who are looking to buy dilutions of odorant compounds: the raw material itself is relatively cheap (twenty euros, or about twenty-seven dollars, for a kilogram of limonene, for example); it is mainly the dilution process that drives up the total cost, which also includes minor expenses associated with packaging, labeling, and compliance with the relevant regulatory standards.

A LEXICON OF BASIC CULINARY ODORS

The art of imparting fragrance, usually (and, in my view, unfortunately) called "aromatization," needs an adequate technical vocabulary. The short list I give in this section is an attempt to group sensations on the basis of what might be thought of as family resemblances, which owe nothing to biology and everything to subjective—and, in many cases, culturally influenced—perceptions. Note-by-note cooks will be wise to make these terms their own, adding to them in case they notice any omissions, because perfumery is something quite different from cooking.

ALDEHYDIC the characteristic smell of heated iron, sea water, fat, sweat

ANIMAL musk, civet, ambergris

BALSAMIC a heavy, sweet odor associated with chocolate, vanilla, cinnamon

CAMPHORATED impregnated with the smells of camphor and similarly scented products

CITRUS a refreshing smell, with hints of lemon and orange

EARTHY the odors of rich, damp soil

FATTY redolent of streaky bacon and lard

FLORAL fragrances evoking the scents of flowers

FRUITY reminiscent of the scents of fruit

GREEN the familiar fragrances of cut grass and leaves

HERBACEOUS a complex odor combining notes of grass, sage, menthol, eucalyptus

MEDICINAL the singular odor of disinfectants (phenolic, cresolic, and the like)

MENTHOLATED fragrances evoking the scent of menthol

METALLIC an odor noticed near metal surfaces

OAKMOSS the scent of undergrowth and seawater

POWDERY reminiscent of the scent of talc

RESINOUS odors of pine sap and other arboreal exudates

SPICY scents of spices

WAXY the odor of candle wax

WOODY the many scents of wood

ODORANT COMPOUNDS

When we try to describe odorant compounds, the question arises whether they should be grouped in terms of their characteristic odors, with the aid of the preceding glossary, or in terms of their molecular structure. The first method has the advantage of generating a series of lists, but these would be cumbersome and endlessly long. The second has the disadvantage of making it necessary to organize compounds according to a rather elaborate system of classification, but this is offset by the advantage of showing the sheer diversity of the world of odors and, perhaps more importantly, of introducing new terms that the cooking of tomorrow will soon make familiar. Besides, the basic principles of classification are easy to understand. Let's get started, then, beginning with the simplest molecules and moving on to more complex ones.

ALIPHATIC COMPOUNDS

Molecules in aliphatic compounds exhibit a straight chain (or "backbone") of bonded carbon atoms to which many hydrogen atoms are attached, along with a few oxygen, nitrogen, and other atoms. There are a number of different types.

- *Hydrocarbons.* These are the simplest organic compounds in a sense because, as their name indicates, they are composed only of carbon and hydrogen atoms. They include gaseous substances used as fuel, such as methane, ethane, propane, and butane. With the addition of more carbon atoms, these compounds are no longer gaseous at ambient temperature and pressure, but liquid or solid as well as inert and odorless.

  Modifications of molecular structure need to occur in order for odors to appear, particularly when neighboring carbon atoms are held together by two chemical bonds rather than by one bond (in which case the compounds are said to be "unsaturated"). Whether hydrocarbons are saturated or unsaturated, linear or branched, they are abundant in edible substances, but few of them contribute significantly to the odor of foods. Examples of odorigenic hydrocarbon compounds include 1,3-trans-5-cis-undecatriene and 1,3-trans-5-trans-undecatriene, which are largely responsible for the smell of galbanum resin (and the oil extracted from it).

- *Alcohols.* Replace a hydrogen atom in a hydrocarbon by a hydroxyl group, in which an oxygen atom is bonded to a hydrogen atom, and you get an alcohol. Saturated alcohols are common in foods, notably in fruits. Since their odor is mild, perfumers do not often use them, but the food industry frequently does in producing compositions and extracts (linear alcohols with four to ten carbon atoms, for example, and isoamyl alcohol). The list of such compounds is extensive. To name just a few:

  3-octanol, a colorless liquid that has an odor of earth and undergrowth found in
  mushrooms, is used to make—you guessed it!—mushroom odors.

  2,6-dimethyl-2-heptanol, a colorless liquid with a delicate scent reminiscent of free-
  sia, a South African iris, has not yet been found in natural substances.

Trans-2-hexen-1-ol, found in many fruits, has a green smell, milder than cis-2-hexen-
1-ol, which it very closely resembles (compounds having the same molecular description but nonetheless different molecular structures are known as isomers).

Cis-3-hexen-1-ol is a colorless liquid with the characteristic smell of freshly cut grass
that accounts for 30 percent of the essential oil of green tea.

1-octen-3-ol, our old friend octenol, passes off in the water vapor of mushrooms. Also
found in the essential oil of lavender, it has an intensely earthy odor, redolent of
undergrowth.

9-decen-1-ol, with a fresh rose smell, is used to make rose-scented soaps.

10-decen-1-ol has a slightly lemony, green, fatty smell and imparts a fresh note to the
strongest floral fragrances.

2-trans-6-cis-nonadien-1-ol, with a smell of violet leaf, is found in cucumber, violet
leaf, and violet blossom oils. The food industry values it for its fresh, green scent
of cucumber.

▪ *Aldehydes and acetals.* The class of aldehydes is molecularly similar to alcohols. Instead of having a hydroxyl group, aldehydes have a single oxygen atom bonded to a carbon atom. Aliphatic aldehydes are essential in
perfumery. Aldehydes whose molecules contain a small number of carbon atoms (acetaldehyde, isobutyraldehyde, isovaleraldehyde, 2-methyl-
butyraldehyde) are used to give fruity and roasted odors. Aldehydes with
between eight and thirteen carbon atoms are widely used in preparing
what I propose be called "odorigenic" (rather than aromatic) extracts and
compositions, or OECs for short. (More than once in the preceding pages
I have criticized the use of the words *aroma* and *aromatic* in connection
with synthetic fragrances. Here, more positively, I am pleased to be able to
recommend a new and more accurate term.) The intensity of the odor of
these compounds diminishes with their molecular size.

Acetals are formed by combining alcohol and aldehyde molecules. Acetals derived from aliphatic aldehydes have characteristic odors that resemble those of aldehydes, though they are less concentrated. They are
frequently used in alcoholic beverages, but seldom in foods because they
are unstable. This class of compounds includes:

Hexanal (more formally, caproaldehyde)—found in apples, strawberries, oranges, and lemons—has a green, fatty odor reminiscent of unripe fruit.

Octanal (caprylalehyde) is naturally present in several citrus oils, notably in the oil extracted from oranges. Its strong odor acquires a citrus note when diluted.

Nonanal (pelargonaldehyde) has a rose smell and is found in citrus and rose oils.

Decanal (capraldehyde) is found in many essential oils and citrus-peel oils. It has a strong odor of orange peel, which is transformed into a fresh citrus odor by means of dilution.

Undecanal (undecyl aldehyde) is also found in citrus essential oils. Its odor is flowery, waxy, fresh.

Dodecanal (lauraldehyde or lauric aldehyde) has a waxy odor that evokes the smell of violet when it is concentrated. Found in citrus fruits and pine needles, it is used to impart citrus notes in OECs.

Tridecanal (tridecanaldehyde), present in lemons and cucumbers, has a waxy, citrus odor and gives fresh top notes.

2-methyldecanal has not been observed in nature. It has an odor of citrus peel.

2-methylundecanal is not yet known to exist in nature either. It has a fatty odor with notes of incense.

Trans-2-hexanal is found in essential oils and the green leaves of many plants. Its grassy odor, sharp and rather pungent, becomes green and pleasant when encountered in more diluted form, recalling the scent of apple.

Cis-4-heptanal is a so-called trace compound (one that can be smelled even though its presence is not detectable by analytical techniques) found in many processed foods. It has a fatty odor, slightly fishy with creamy nuances.

10-undecanal has an odor that is at once flowery, heavy fatty, green, slightly metallic.

◘ *Ketones.* Members of this class of aliphatic compounds have quite different odors from the ones already cited. Ketones have an oxygen atom double-bonded to a carbon atom in the middle of the chain, not to a terminal carbon. I will content myself with mentioning but two examples from a long list:

3-hydroxy-2-butanone (acetoin) has a buttery odor used to add flavor to margarines.

2,3-butanedione (diacetyl), present in many fruits and in butter, is added to both processed butters and margarines.

■ *Acids and esters.* Let us conclude this brief survey by looking at carboxylic acids and at esters, the compounds these acids form when they react with alcohols. Carboxylic acids are seldom used in either perfumery or the OEC sector of the food industry, but esters (especially acetates, which is to say esters containing acetic acid) are commonly used in both to enhance fruit odors. These compounds include:

Ethyl formate, an ester of formic acid (so called because it is present in the urticating liquid released by many kinds of ants), has a fruity, ethereal, rather assertive odor.

Cis-3-hexenyl formate has been identified in tea. Its green and fruity odor is used to impart green notes to a variety of commercial preparations.

Ethyl acetate has a fruity odor and is found in many kinds of fruits.

Butyl acetate has a strong fruity odor of the sort that is especially pronounced in apples.

Isoamyl acetate is the compound principally responsible for the odor of bananas.

Hexyl acetate has a smell evoking that of peaches.

3,5,5-trimethylhexyl acetate, not found in nature, has a woody scent.

Trans-2-hexenyl acetate, present in fruits and mint essential oil, has a fruity, green odor.

Ethyl propionate, found in fruits, has an odor reminiscent of rum.

Ethyl butyrate, present in fruits and cheese, has a fruity odor recalling that of grapefruit.

Butyl butyrate is present in fruits and honey and contributes to their odors.

Isoamyl butyrate, which has a strong fruit odor, occurs in bananas.

Hexyl butyrate likewise has a strong fruit odor.

Cis-3-hexenyl isobutyrate, present in mint oil, has a fruity, green scent.

Ethyl isovalerate has an odor of blueberry.

Ethyl 2-methylbutyrate, present in citrus fruits, has a green odor redolent of apple.

Ethyl hexanoate and 2-propenyl hexanoate have an odor of grapefruit.

Ethyl heptanoate has scents of cognac.

Ethyl octanoate has a floral, fruity odor.

Ethyl 2-trans-4-cis-decadienoate, identified in pears, has an odor reminiscent of pear brandy.

ACYCLIC TERPENES

By now, and probably not for the first time, it will have become clear to you that the number of fresh opportunities for culinary creativity presented by compounds is simply staggering. What the note-by-note cook must keep on reminding himself is that I have so far only scratched the surface of the world of odorant compounds. I have not yet even mentioned terpenes, whose immense diversity is associated with the fact that their basic molecular structure is repeatedly encountered in the multifaceted process of plant synthesis. This time, rather than scroll through a long list, I give only a few examples of compounds that are already well known in the world of perfumery:

Geraniol (3,7-dimethyl-trans-2,6-octadien-1-ol), found in many essential oils, is a colorless liquid with a flowery, slightly soapy odor. In odorigenic compositions, a small quantity accentuates the citrus note.

Nerol (3,7-dimethyl-cis-2,6-octadien-1-ol) is generally found in the company of geraniol. It, too, is a colorless liquid, but with a fresher rose scent.

Linalool (3,7-dimethyl-1,6-octadien-3-ol) is likewise present in many essential oils. It is the principal constituent (60–70 percent) of coriander essential oil, for example. It has a fresh odor, redolent of lily, and is highly volatile, giving a "natural" quality to top notes.

Myrcenol (2-methyl-6-methylene-7-octen-2-ol) has been isolated in essential oils. It accentuates notes of citrus and lavender.

Citronellol (3,7-dimethyl-6-octen-1-ol), usually produced by mixing together two enantiomers, is a colorless liquid with a soft rose odor.

Citral (3,7-dimethyl-2,6-octadien-1-al), one of the most widely used of all odorant compounds, is typically a mixture of two isomers that appears as a nearly colorless liquid, slightly yellowish, with a lemony odor. It is found in citronella oil and is used in a great variety of citrus compositions.

CYCLIC TERPENES

Up to this point I have spoken of compounds whose chains of carbon atoms are linear or branched, but not cyclic (or ring shaped). I now turn to cyclic

compounds for several reasons, not least because of the occasion they provide to pay tribute to the English physicist and chemist Michael Faraday
(1791–1867), who discovered benzene, regarded today as the prototype "aromatic" compound. The soft, agreeable smell of benzene (a carcinogen, as
it happens) is a molecular characteristic determined by the bonds among its
carbon atoms—yet another reason not to speak of "aroma" in note-by-note
cooking. We will avoid confusion by speaking of "odor" instead.

The basic structure of cyclic compounds is quite simple. If you take a
chain of carbon atoms and then loop it back on itself so that the atoms at each
extremity are now attached, you obtain what chemists call a ring. Cyclic compounds may contain rings composed of three, four, five, six, or more carbon
atoms. Some may also include oxygen, nitrogen, and other atoms.

Certain terpenes, found in many essential oils, are cyclic. Limonene
(1,8-p-menthadiene), a liquid with a lemony odor, occurs in citrus peels in the
form of its (+) isomer in concentrations as high as 90 percent; the (−) isomer
is found in mints and conifers. Menthol exists in several forms: (−)-menthol,
(+)-neomenthol, (+)-isomenthol, and (+)-neoisomenthol, the first being the
most frequent; the principal compound of mint essential oils, menthol is also
refreshing. Carvone (1,8-p-menthadiene-6-one) also exists in several liquid
forms, all of them slightly yellowish, but with very different odors: in the (+)
form, carvone is the principal compound of caraway and dill essential oils;
in the (−) form, it is present in mint. Fenchone (1,3,3-trimethylbicyclo[2.2.1]
heptan-2-one) is a compound whose (−) isomer has an odor of fennel. Ionones, produced by the degradation of carotenoids—yellow, orange, and red
pigments that are cousins of the beta-carotene found in carrots—are present
in many essential oils.

OTHER AROMATIC COMPOUNDS

Not all cyclic compounds are terpenes, of course. Other aromatic compounds—I would be happier, as I say, if they were called odorigenic compounds—are cyclic as well. For example, P-cymene is present in many essential oils; in its pure form, it has a citrus scent. Benzaldehyde has a bitter
almond odor. Cinnamaldehyde (3-phenyl-2-propenal), a somewhat spicy

liquid in which the trans isomer is dominant, is the principal compound of cinnamon.

PHENOLS AND PHENOLIC DERIVATIVES

When aromatic compounds contain a hydroxyl group, they become phenolic compounds, a class of organic compounds in which a hydroxyl group is bound directly to a carbon atom in a benzene ring. This class includes the following compounds:

Anethole (1-methoxy-4-[1-propenyl]benzene) composes 80 to 90 percent of the essential oil of anise, star anise, and fennel. It is this molecule, soluble in alcohol but insoluble in water, that causes the cloudiness one observes when water is poured into pastis. (The proof that water is dangerous, by the way—as Alphonse Allais, inventor of the frosted-glass aquarium for shy goldfish, famously remarked—is that a single drop is enough to cloud the purest pastis.

Eugenol (2-methoxy-4-allylphenol), also a yellowish liquid with a spicy odor, is the principal constituent of several essential oils, notably clove and cinnamon.

Isoeugenol (2-methoxy-4-[1-propenyl]phenol), a viscous, yellowish liquid with an odor of clove, is often found together with eugenol in essential oils.

Thymol (2-isopropyl-5-methylphenol), the principal constituent of thyme and oregano essential oils, forms colorless crystals.

Vanillin (4-hydroxy-3-methoxylbenzaldehyde) is a colorless solid with a vanilla odor and the principal constituent of vanilla essential oil.

4-(4-hydroxyphenyl)-2-butanone has a strong raspberry odor.

O-, N-, S-HETEROCYCLES

I spoke earlier of cyclic compounds in which a carbon atom is replaced by an oxygen, nitrogen, or sulfur atom. By way of conclusion, here are a few examples that I hope will inspire note-by-note cooks to use heterocycles, as these cyclic compounds are called, albeit in suitably diluted form:

2-furaldehyde has an odor of freshly baked bread.

2-methylfuran-3-thiol has an odor of roasted beef.

Eucalyptol (1,8-epoxy-ᴘ-menthane) is used to impart a note of freshness.

Maltol (3-hydroxy-2-methyl-4H-pyran-4-one) occurs in the form of colorless needles
having an odor of caramel or freshly baked cake. It is used as a flavor enhancer in
fruity compositions, especially ones in which a note of strawberry is predominant.

Gamma-decalactone has a peach odor. Delta-decalactone is used for creams and
butters.

ON THE PROPER USE OF ODORIGENIC
EXTRACTS AND COMPOSITIONS

Nature does, of course, provide us with an abundance of fragrant products
that are easy to use—herbs and spices, for example—but these things do not
always make a very powerful impression, for the odorant compounds they
contain have to migrate from the ingredient to the food before then passing
more or less completely into the mouth, as we saw in connection with tastes
(recall the matrix effects discussed in chapter 3). OECs, by contrast, so long
as they are dispersed on the surface of foods, make their effects felt at once
and without limitation. What's more, such preparations do not spoil, nor are
they sensitive to the vagaries of the weather. The cook who hopes to experiment successfully with these products will nevertheless need to keep in mind
three essential points: he must choose a substance that is well suited to his
purpose, incorporate it in his dish at the right moment, and use it in the right
amount.

With regard to proper dosage, first, the note-by-note cook will profit by
learning from perfumers, who take care to distinguish between top notes,
the first notes to be sensed; middle notes, the next ones to be perceived; and
bass notes, which are recognized last. Top notes are often derived from compounds with the smallest and most volatile molecules, but both molecular
structure and the environment into which compounds are introduced need
to be reckoned with. A compound is released more rapidly when it is pure
than when it is dissolved in a fatty substance, for example.

The art of perfumery can also teach cooks that there must be a kind of bal-
ance, or equilibrium, between top and middle notes. Aromaticians—counter-

parts to perfumers in the food industry, who specialize in OECs—are keenly aware that there must be a more general equilibrium between odor and taste as well. Aerated products, for example, serve to make top notes more prominent: the longer a foam can be kept in a stable state before being consumed, the more time there is for the leading compounds to pass into the air inside the bubbles that make up the foam before being released in the mouth, where they are immediately recognized by the organs of taste and smell.

Basic physical chemistry suggests many other rules of thumb. The cook must, for instance, take into account the absorbent properties of gums and other fatty substances that constitute the chemical environment of an odorant compound: the more thoroughly the compound is absorbed, the less of it is released; conversely, the greater the absorption, the longer it takes for the compound to be precipitated (or evolved as a gas). This is why some products (recall once again the matrix effects we examined in connection with taste) accentuate the perception of certain compounds, whereas others make them seem less assertive. Milk, for example, has a powerful masking effect, unlike alcohol, which heightens the taste and smell of foods, giving them color and sparkle. The temperature at which a dish is consumed is no less important, as makers of processed pork products well know. The same thing is true in the case of odor, for heating causes the most volatile compounds to evaporate. A food eaten cold will therefore have to contain a larger quantity of odorant compounds than the same food eaten hot if it is to have a comparable sensory effect.

TRIGEMINAL SENSATIONS

There is an important but so far poorly understood class of sensations whose receptors, like the ones responsible for our perception of odors and tastes, are located in the nose and the mouth but are not therefore reducible to odors and tastes. They are associated with a specific neural pathway, the trigeminal nerve, which runs down from the back of the skull and divides into three branches—hence the nerve's name. The receptors at the ends of these branches register both fresh, cooling sensations and the sensations of pungency and heat produced by many spices. These receptors seem also to play an important role in the perception of pain, although in this connection their functioning is even less well understood. There's no point belaboring the fact of our ignorance. We must step aside and let science get on with its work.

The food industry spends a great deal of time and money investigating the properties of coolness and heat. Commercial secrecy is intense. Manufacturers of compounds and preparations that stimulate trigeminal sensations jealously guard the results of their research, and still today note-by-note cooks do not have ready access to information of the sort I have just given regarding odorant compounds. Here and there I have mentioned in passing a certain freshening compound, such as menthol (extracted from mint), eugenol (from clove), and xylitol (from various fruits). The hard work of discovering exactly how their effects are produced nevertheless remains to be done—and not only by scientists. A step forward would be taken, for example, if an online discussion forum were to be set up where cooks could post tasting notes about refreshing and pungent ingredients. Head colds are not a disability in this enterprise, by the way, because they allow distinctions to be made that will be of value to sensory physiologists. The next time you come down with a cold, taste some mustard, a hot pepper, some ground black pepper, watercress, a raw chanterelle mushroom, some raw garlic, and then try to describe the sensations. Why these particular ingredients? Because they all are hot or pungent in one way or another. Do the same thing with ingredients that impart a cooling, refreshing sensation (or at least are popularly supposed to do so), once again keeping in mind that some people perceive food sensations mainly through the mouth, others mainly through the nose; and also that the trigeminal system, though it is not reducible to tastes and odors, is not unrelated to them either, for a compound that is cool or hot often has a distinctive odor or taste, or both.

Many people have heard it said that the pungency of black pepper is due to piperine and that the pungency of hot peppers is due to capsaicin. And so they are. But these compounds are only two among a great many more or less similar compounds that have trigeminal effects. Recent studies of vanilla have shown that the pods also contain, in addition to the vanillin compound I mentioned earlier, velvety-tasting compounds. Are these compounds associated with a presently unknown trigeminal sensation? No one knows. But chemists have long been aware that the world of natural products—the good Lord's rich storehouse, as it is referred to in the Bible—is far more vast than the part of it that is known to us today. It is high time that cooks were aware of this as well.

FURTHER READING

Borysik, A. J., L. Briand, A. J. Taylor, and D. J. Scott. "Rapid Odorant Release in Mammalian Odour Binding Proteins Facilitates Their Temporal Coupling to Odorant Signals." *Journal of Molecular Biology* 404, no. 3 (2010): 372–380.

Calvino, B., and M. Conrat. "Pourquoi le piment brûle." *Pour la science* 366, no. 4 (2008): 54–61.

Charles, M., S. Lambert, P. Brondeur, J. Courthaudon, and E. Guichard. "Influence of Formulation and Structure of an Oil-in-Water Emulsion on Flavor Release." In D. D. Roberts and A. J. Taylor, eds., *Flavor Release*, 342–354. Washington, D.C.: American Chemical Society, 2000.

Labbe, D., F. Gilbert, N. Antille, and N. Martin. "Sensory Determinants of Refreshing." *Food Quality and Preference* 20, no. 2 (2009): 100–109.

Surburg, Horst, and Johannes Panten. *Common Fragrance and Flavor Materials: Preparation, Properties, and Uses*. 5th ed., completely revised and enlarged. Weinheim, Germany: Wiley-VCH, 2006.

FIVE COLOR

FOOD DYES ARE OFTEN disparaged today, but, as Cicero tells us, a man who knows only his own generation remains always a child. The history of cooking shows that coloring agents have been popular since the earliest times. In the Middle Ages, cooks used a variety of substances derived from spices and vegetables, even insects. Green was obligatory in the Christian West, where it symbolized the resurrection of Jesus Christ. Modern cooks continue to use the green pigment of spinach to color sauces, for example. Nothing could be simpler: grind up some spinach leaves in a blender, press out the juice through a cheesecloth-lined sieve, and then simmer over low heat until a very green froth—the dye itself—bubbles up from the brownish liquid.

THE EYE PRECEDES THE PALATE

Achieving mastery over the outward appearance of foods products is an ancient ambition because it has long been known that the faculty of taste can be fooled by the visual aspect of a dish. Chefs today continue to be intrigued by the possibility of using first impressions in order to influence the judgment of diners. Physiologists, for their part, have thoroughly investigated how the brain processes the information it receives about foods once they have been swallowed. The sensations detected by taste and olfactory receptors are encoded as neural signals in a dense and rapid train of electrochemical impulses. Only certain nerve fibers are excited in reaction to the detection of a particular flavor. Accordingly, a mapping of fiber-excitation patterns reflects the coexistence of various flavors. For each of our senses, a cerebral area of a few square centimeters registers a neural image of the sensation that has been transmitted to the brain for interpretation. First, this image is presented to memory, which reports back that it does or does not recognize the image or that the image resembles another previously stored image. Then the brain integrates all the relevant sensory and cognitive data, determines the level of pleasure aroused by the sensation, and commits this new piece of information to memory, where it forms part of the correspondingly enlarged archive that will be consulted in assessing the next sensation. In other words, we are guided in our judgment of a dish by the preliminary estimates made by the brain—in the same way that we are pleased once again to see someone we like because our brain associates seeing this person with a pleasurable response. This fact has a crucial implication, that our first impression is right (or at least not entirely mistaken)—which implies, in turn, that vision is an important aspect of our perception of flavor.

Experimental confirmation was strikingly provided a few years ago by researchers at the Institute of Oenology at the University of Bordeaux. Subjects trained in the use of wine-tasting terms (so-called descriptors that characterize the aroma, or bouquet, of red and white wines) were asked to describe two samples: a simple white wine that was made from a mixture of grapes and displayed no distinctive varietal character (or "typicity"), and the same wine colored red using pigments extracted from red wine. None of this was known to the subjects, who proceeded in all innocence to describe the white wine with the vocabulary of white wine and the faux red wine with the vocabulary of red wine! The tasters' response is not as surprising as it seems. The eye is preeminent among human sense organs: visual information is the first type of sensory information to be processed since it is the first to reach the brain. Once a neural image has been received, the brain seeks to specify the conceptual category to which it belongs by consulting its archive of gustatory information. If it is fooled in the first place by the color of a substance, it has a hard time finding its way out from the category into which the true perception of this color initially led it.

LEGALLY APPROVED COLORING AGENTS

Cooks are free, of course, to go to the trouble of making green dye from spinach and other natural foods, but why should they deprive themselves of coloring agents that food manufacturers are legally authorized to use? The list of additives formally approved for use in the European Union and Switzerland contains an impressive variety of products bearing E numbers in the 100s. Let's take a brief tour, noting a few details along the way that may be of interest to cooks wishing to experiment with products they have never tried before.

E100 Curcumin is a yellow dye found in the rhizome of a plant of the ginger (Zingiberaceae) family, *Curcuma longa,* which contains turmeric acid, the substance that gives curry powders their pungency. Obviously curcumin is not to be confused with the spice itself, turmeric. It has the advantage of imparting color without also imposing the powerful trigeminal sensation of the spice. Should turmeric in its powdered form nevertheless be preferred to E100? There are two reasons for doubting it. First, the powder is obtained by grinding and drying, operations causing chemical degradations that make the powder less pure than the liquid extract. (Just so, it would be a welcome thing in my view if we were to stop calling turmeric powder "turmeric," for the plant and the powder are very different in their molecular structure.) Second, the powder is much more liable to adulteration than E100, whose manufacture is strictly controlled.

E101 Riboflavin and riboflavin-5′-phosphate are yellow dyes extracted chiefly from yeasts, wheat germ, eggs, and animal livers. Their chemical synthesis is understood as well.

E102 Tartrazine is a synthetic yellow coloring agent. Food manufacturers sometimes mix it with Brilliant Blue (E133) or with Green (E142) to obtain various shades of green, particularly in order to add color to canned baby peas. One hears it said still today that tartrazine causes the testicles to shrivel and reduces sperm count, but as long ago as 1917 it was shown to carry no such risk—another sign of the irrationality to which we are prone in the face of perceived dangers to our health.

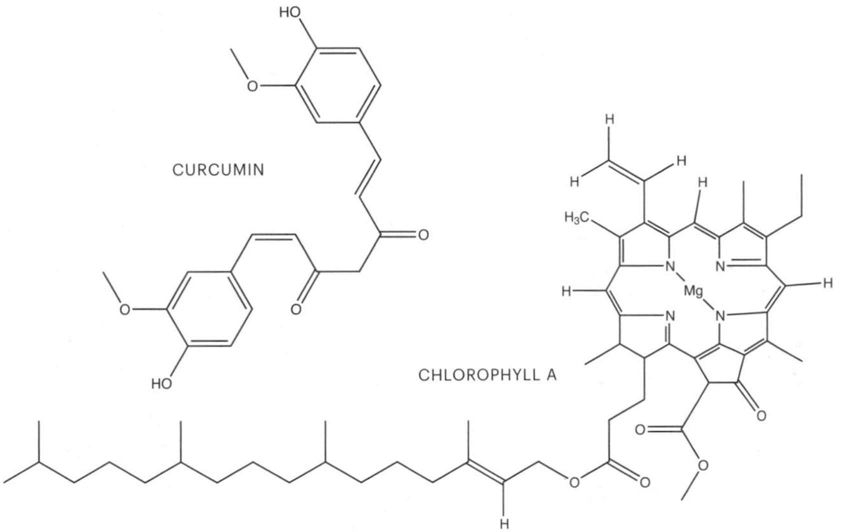

CURCUMIN

CHLOROPHYLL A

5.1 *VEGETABLE MATTER IS USED FOR THE PRODUCTION OF FOOD ADDITIVES. THE RHIZOME OF THE PLANT CURCUMA LONGA, FOR EXAMPLE, IS USED TO PRODUCE CURCUMIN (E100), AND ALFALFA (KNOWN IN EUROPE AS LUCERNE) IS USED TO ISOLATE CHLOROPHYLLS (E460).*

E104 Quinoline (or Quinoline Yellow WS, to give it its official name) is a synthetic yellow dye of a subtly different shade than its neighbors.

E110 Sunset Yellow (Orange Yellow S) is a synthetic orange dye.

E120 This code designates a trio of natural red dyes, cochineal, carminic acid, and carmine. They are prepared from aqueous, aqueous-alcoholic, and alcoholic extracts of the dried shell of the female cochineal insect (*Dactylopius coccus Costa*)—found in Central America, the Canary Islands, North Africa, and southern Spain—from which the color takes its name. The shell encloses eggs and young larvae. About fifteen thousand of these insects are needed to obtain a hundred grams of dye. The principal pigment is carminic acid.

E122 Carmoisine (or azorubine) is a synthetic red dye discovered about a century ago. First used (along with cochineal) by textile manufacturers, it was subsequently adopted by printers and later by the food industry.

Should the commercial migration of dyes from cloth production to edible products make us uneasy? The answer to all such questions depends on how a given compound interacts with the human body. Is it metabolized? If so, how? What chemical products are formed by the metabolic reactions? Do these products have receptors? In what quantities do they have an effect on the organism? Regulatory standards specify approved uses for additives. The use of carmoisine, for example, is approved in Europe (but not in the United States) for pork sausage casings as well as for preserved fruits, candies, ice creams, and beverages. Industrial applications are carefully monitored on the whole, but the artisanal use of such products often goes unnoticed. Titanium dioxide, for example, is a white dye approved for sausage casings, cheese rinds, decorative sugar work, and chewing gums, but I have seen more than one pastry chef use it for other purposes without anyone objecting. Evidence of a double standard?

E123 Amaranthe is a red synthetic dye. Some thirty years ago it became a source of toxicological controversy even though many studies had shown it to be harmless. Without going into the details of this episode, let me say simply that commercially manufactured additives are required to indicate on their labels an acceptable daily intake (ADI), which is based on a calculation known as LD50: the letters L and D stand respectively for "lethal" and "dose"; the number 50 means that this dose is sufficient to kill half the members of a test group of laboratory animals. In determining the ADI, toxicologists take great care to multiply the LD50 value by a considerable factor in order to offset the risks posed by worst-case scenarios.

Note-by-note cooks must be sure to inform themselves about all the precautions needing to be taken whenever they consider using a compound, either by researching the matter themselves or by consulting their supplier. To give you some idea of the proportions that are typically involved, the ADI for carmoisine is 0.8 milligrams per kilogram of body weight (about 0.36 milligrams per pound). That doesn't sound like much, but it has to be kept in mind that carmoisine in its pure form is an extremely powerful staining agent. It should be used in minimal quantities.

E124 Ponceau 4R (also called Cochineal Red A) is a synthetic red dye that has nothing to do with the natural dye called cochineal that I mentioned earlier.

E127 Erythrosine is a red food coloring. As in the case of some other dyes, commercial use of its sodium, calcium, and potassium salts is also permitted. These salts are designated by the same code, supplemented by the letters *a*, *b*, *c*, and so on (or by the numerals *i*, *ii*, *iii*, and so on) as necessary.

E128 Red 2G (or azogeranine) is, as the name suggests, yet another red coloring agent. (I forgot to mention, by the way, that the dyes discussed here are for the most part powders, granules, or concentrated solutions.)

E129 Allura Red AC. Red, again—only different.

E131 Patent V Blue. A synthetic blue dye.

E132 Indigo carmine (or indigotine). In France it was a pigment of this type—used to make blue jeans (*toiles de Nîmes*, source of the English word *denim*) blue—that helped bring wealth and prosperity to Toulouse and the neighboring area of Cocagne (between Toulouse, Castres, and Albi).

E133 Brilliant Blue, the blue dye I mentioned earlier in connection with tartrazine.

E140 Chlorophylls and chlorophyllins. Let's pause here for a moment, for these pigments contribute very largely to the green color we see in algae and plants. There are many chlorophylls, and every vegetable contains several of them (which means that the term *chlorophyll*, in the singular, should be avoided because it suggests there is only one kind). But they are not solely responsible for the color of vegetables. Nor are all of them green. The green we see in spinach, for example, is due to a combination of pigments in which several chlorophylls (yellow green, green, green blue) are mixed with carotenoids (yellow, orange, and red). Painters are therefore right to add red to green when they draw plants, for this in fact is how plants make their green colors—something that cooks who wish experiment with colors need to remember.

The additives jointly designated by the code E140 are produced by extraction from herbs, alfalfa, nettles, and other edible vegetable matter. Eliminating the solvent can lead to total or partial separation of the magnesium found in chlorophyll molecules, which changes the color somewhat. All chlorophylls have a plate-shaped molecule, more or less square with a sort of tail attached to one corner and a magnesium atom at the center. When a green vegetable is heated, especially in an acidic medium, the magnesium is expelled, and the chlorophylls are transformed into pheophytins, which have a bluish tint. This phenomenon is well known to cooks, who counteract it by adding baking soda to the cooking water to prevent the loss of magnesium and to retain the vibrant green color of fresh vegetables.

E141 Copper complexes of chlorophylls and chlorophyllins. These green pigments were first isolated as part of an attempt to explain why the characteristically vivid color of vegetables disappeared during prolonged cooking. It had been observed that vegetables kept their color when they were cooked in copper pans. Cooks, not realizing that the copper served to replace the lost magnesium at the center of the chlorophyll molecules, therefore became fond of using what they called "greening pans." (For a certain time, as I mentioned earlier, it was even customary to add copper sulfate to the cooking water—a procedure that was subsequently found to pose unacceptable risks and was prohibited by law in France in 1902.) In such matters we should not be reluctant to place our confidence in government agencies: if the use of copper complexes of chlorophylls is permitted today, it is because exhaustive toxological testing over many years has shown it to be safe.

E142 Green S. Yet another green dye.

E150a Caramel. That's right, ordinary caramel, used as a brown coloring. Generally speaking, compounds containing brown and black pigments have an odor of burnt sugar and an agreeable, slightly bitter taste. But how has plain caramel wound up on a list of color additives? Read on.

E150b Caustic sulfite caramel. Plain (let it be noted, not "natural") caramel is obtained by cooking sucrose over high heat. It is a complex mixture: some of its constituent compounds impart consistency, others taste, others odor, and others still color. Cooks have long been aware that caramels differ depending on the conditions under which they are made. Syrups to which a dash of vinegar has been added, for example, caramelize differently than the same syrups containing no acid. Adding other ingredients yields other results. And so it is that the caramel identified by the code 150b comes by its name.

E150c Ammonia caramel, due to the addition of ammonia.

E150d Sulfite ammonia caramel. Oh, while I'm thinking of it: a few years ago the Italian Ministry of Health banned the use of coloring additives in restaurant kitchens as part of a demagogic campaign against molecular cooking. The additives in question? Caramel and its derivatives, such as sulfite ammonia caramel. Revolutions have been started for less than that!

E151 Black PN (or Brilliant Black BN). Yes, a black dye.

E153 Vegetable carbon. A black pigment produced by carbonizing vegetable matter such as wood, cellulose residues, peat, coconut husks, and the outer covering of other vegetables. The raw material is charred at high temperatures.

E154 Brown FK, a brown dye.

E155 Brown HT, yet another brown dye.

E160a Alpha-carotene, beta-carotene, gamma-carotene. Now we come to the carotenoids, pigments present in many vegetables and the source of the brilliant yellow, orange, and red colors of many edible fruits (lemons, peaches, apricots, strawberries, cherries, and so on), vegetables (carrots, tomatoes, and so on), mushrooms (chanterelles), and flowers. They are also present in a variety of animal products, including eggs, lobsters, langoustines, and fish. I won't go into the molecular details except to note that heating may modify their color and that they can react with different compounds. Beta-carotene, abundant in carrots, is perhaps the best known of the three pigments designated by this code.

E160b Annatto, bixin, norbixin. Annatto is a seed that is very commonly used in South America, among other places. Some indigenous peoples in the Amazon, for example, use it to paint their skin red. Bixin is prepared by extracting annatto seeds from their husks with the aid of solvents, and hydrolyzed to norbixin by using an alkaline solution. Annatto extracts may be either aqueous or oily.

E160c Paprika oleoresin (paprika extract), capsanthin, capsorubin. The extract, obtained by the action of a solvent on natural strains of paprika, contains paprika's principal pigments, capsanthin and capsorubin.

E160d Lycopene. A pigment naturally present in tomatoes and many other vegetables and fruits. It's fun to make it yourself, but easier to use when it comes already made in a bottle.

E160e Beta-apo-8'-carotenal. Another red pigment.

E160f Ethyl ester of beta-apo-8'-carotenic acid. An orange-red to yellow pigment.

E161 This designation covers ten compounds, molecular neighbors of the carotenoids and known as xanthophylls. They are found in many vegetable substances and include the next two coloring agents.

E161b Lutein. A xanthophyll prepared by solvent extraction from edible fruits and plants as well as from herbs, in particular alfalfa (lucerne).

E161g Canthaxanthin. A cousin to lutein, canthaxanthin assumes the form of beautiful dark-violet crystals.

E162 Beetroot Red. More formally known as betanin, it is obtained from natural strains of red beetroot either by pressing crushed beetroot until it yields a liquid or by aqueous extraction from shredded beetroot, followed by enrichment of the betacyanins that constitute the main coloring principle.

E163 Anthocyanins—a world of their own, for they comprise the coloring substances of fruits and flowers. Take a rose, grind it up with water, and strain the liquid, and you are left with an impure solution of anthocyanins, the pigments of the rose. If you acidify this solution by adding a drop of white vinegar, you see that the color changes. Now add a little baking soda to raise the pH (by lowering the acidity), and you get another color. It is generally true of anthocyanins that their color depends on the acidity of the medium and on the presence of metal ions. If you add a little iron (found in eggs and meat) or a little zinc or a little aluminum, the colors completely change. Indeed, the entire range of shades can be recreated, from yellow to violet.

Most anthocycanins used as food dyes are prepared by extraction from edible vegetable tissues using sulfited water, acidified water, carbon dioxide, methanol, or ethanol. The pigments themselves contain organic acids, sugars, mineral salts, and tannins. Tannins are an especially interesting kind of phenolic compound, by the way, because they are neither sweet nor sour nor bitter nor salty; they have no odor, nor are they cool or pungent. They are, well, tannic—which means that in binding with proteins in the saliva, they suppress the lubricating action of these proteins, causing the soft tissues in the mouth to feel as though they have been constricted. When the astringent effect is strong, it may be disagreeable, but when it is slight, as in the case of certain grapes, the sensation is marvelous. Conveniently enough, the winemaking industry sells a whole host of "oenological" tannins for the note-by-note cook to play with!

Some colorants are authorized for external use only:

E170 calcium carbonate (chalk)

E171 titanium dioxide (mentioned earlier)

E172 iron oxides and iron hydroxides

E173 aluminum

E174 silver

E175 gold

E180 lithol rubin BK (a reddish synthetic powder approved for staining cheese rinds)

I should emphasize that there is no reason why these pigments must be used in their pure form. By mixing them, note-by-note cooks will be able to produce a range of almost unimaginably brilliant colors and shades, sharper and more vivid than the ones with which they are familiar from traditional cooking. Recipes for new dishes will need to be developed, of course, but this will be scarcely more difficult than learning to paint with gouache—child's play really, a game that takes its place alongside the ones we learned to play earlier involving shape, consistency, taste, and odor.

NATURAL VERSUS ARTIFICIAL REDUX

The idea of using artificial food dyes has a way of crystallizing unfounded fears and strengthening popular opposition to "chemistry." The most recent charge is that these products are responsible for hyperactivity in children. Even if hyperactivity were to be proved, however, would it not more likely be the result of kids spending too much time in front of television and computer screens? However this may be, fearmongers are the scourge of the modern world. Some of them are honest, others are not; some of them are naive, others are not. All of them are opposed by trained specialists who reject their opinions—alas, to no avail. If people no longer trust experts or politicians or government agencies or the press, can the day be far off when even our academies of science no longer command respect?

For many years now, much of the public has believed that synthetic colorants should be banned because they are "bad," but that natural colorants are "good," especially if they have been prepared without an organic solvent. Without an organic solvent? Ethanol, of which we are so fond, is an organic solvent! Our inconsistency on this point is really quite troubling. We want to have colors—without coloring agents. And yet several years ago, when makers of a certain mint syrup stopped using the green dye they had always added (mint syrup is colorless, as it happens), the clamor for its return was deafening. Arguing over whether natural green should be used rather than synthetic green is utterly beside the point: we have seen more than once that what is natural is not always good and what is artificial is not always bad— quite to the contrary.

In making sense of the claim that natural colorants are inherently superior to artificial colorants, it will be instructive to consider a pigment that French scientists at the National Institute for Agricultural Research laboratory in Rennes recently discovered. They succeeded in isolating a new water-soluble yellow dye, POP2 (POP stands for "phloridzyn oxidation product"), that can be used as a natural substitute for tartrazine (first synthesized and approved for commercial use almost a century ago). In effect, POP2 is a precious waste product. Ten million hectoliters (almost 265 million gallons) of apple juice are pressed from 45 million tons of apples in the world each year, which makes a massive quantity of solid-waste material available for productive uses of one kind or another. Until recently the food industry had been content to extract pectins from this waste material, which are widely used, as we have seen, in manufacturing fruit preserves, fibers, and animal feed. But the pulp of apples also contains phenolic compounds, all of whose molecules contain at least one benzene group (six carbon atoms arranged in the form of a hexagon) and one hydroxyl group (an oxygen atom bonded to a hydrogen atom). One of these molecules, phloridzin, specifically associated with the Rosaceae family (and especially apples), happens to be the precursor of POP2.

Phloridzin is chiefly present in seeds, which means that pressing apples concentrates it. The conversion of this precursor into a pigment results from a reaction that, under other circumstances, would be considered harmful. Phenolic compounds are responsible for many foods turning brown once they have been cut up: the knife releases these molecules, along with enzymes known as polyphenol oxidases, which in intact cells are found in separate compartments. In the presence of air, the enzymes transform the phenolic compounds into reactive quinones, which subsequently form dark compounds. This is why apples, avocados, pears, and indeed the majority of fruits and vegetables turn brown.

Cooks and food engineers make a special point of blocking this reaction. Cooks use lemon juice, whose ascorbic acid (vitamin C) chemically lowers the concentration of quinones. Food engineers have a number of methods at their disposal, both physical (centrifugation, which eliminates browned liquids) and chemical (various antioxidants). Now, it is exactly these normally harmful enzymatic oxidation reactions that produce POP2, which turns out to be yellow rather than brown: chemists studying the phenolic compounds found in apples observed that the color of the pulp changed during storage. Using a separation process known as liquid-phase chromatography (in which a mixture is dissolved in a fluid and injected into a column filled with a pulverulent solid so that the various compounds of the mixture will migrate at different speeds), it became clear that three new phenolic compounds are formed. The first one is a derivative in which a hydrogen atom is replaced by a hydroxyl group. This intermediary is then transformed into another colorless compound, POP1, which subsequently is transformed into POP2, an intense yellow pigment.

The reaction that produces POP2 is remarkable for three reasons. First, it is the apple that furnishes both the precursor, phloridzin, and the enzyme. Next, although the reaction occurs more rapidly with increasing temperature, its yield is the same at 10° or 30° or 40°C (50° or 86° or 104°F), so there is no point expending additional energy to heat the material more quickly. Finally, the POP1 that is produced along with POP2 is a good antioxidant (and may come to have a related application one day); POP2, however, is water soluble like many phenolic compounds and in low concentrations produces a saturated color varying from brilliant yellow (between pH 3 and pH 5) to orange for less acidic media. Will POP2 manage to establish itself in the market? A commercial patent has been granted, but the substance itself will nevertheless have to pass toxicological tests, just as synthetic compounds do, and no one knows in advance whether it will turn out to be safe or unsafe. The moral of the story is that in matters of toxicology we are well advised to adopt an impartial attitude—whether the products being tested are natural or synthetic.

FURTHER READING

Davidson, Alan. *Oxford Companion to Food*. Oxford: Oxford University Press, 1999. See especially the entries on food additives and coloring.

Food Additives in Europe 2000: Status of Safety Assessments of Food Additives Presently Permitted in the EU. Copenhagen: Nordic Council of Ministers, 2002.

Guernevé, C., P. Sanoner, J. F. Drilleau, and S. Guyot. "New Products Obtained by Enzymatic Oxidation of Phloridzin." *Tetrahedron Letters* 45 (2004): 6673–6677.

Guyot, S., S. Serrand, J. M. Le Quéré, P. Sanoner, and C. M. G. C. Renard. "Enzymatic Synthesis and Physiochemical Characterisation of Phloridzin Oxidation Products (POP), a New Water-Soluble Yellow Dye Deriving from Apple." *Innovative Food Science & Emerging Technologies* 8 (2007): 443–450.

Milgrom, Lionel R. *The Colours of Life: An Introduction to the Chemistry of Porphyrins and Related Compounds*. Oxford: Oxford University Press, 1997.

Rossotti, Hazel. *Colour: Why the World Isn't Grey*. Princeton, N.J.: Princeton University Press, 1985.

Scotter, M. J. "Methods for the Determination of EU-Permitted Added Natural Colours in Foods: A Review." *Food Additives & Contaminants* 28, no. 5 (2011): 527–596.

Serrand,S., S. Bernillon, J. M. Le Quéré, P. Sanoner, and S. Guyot. "The Role of Polyphenoloxidase in the Synthesis of a Yellow Pigment Derived from Apple." In *Proceedings of Enzymes for Food—Symposium européen, Rennes, France (2006)*, 131–139. Paris: Institut National de la Recherche Agronomique, 2006.

This, Hervé. "Artificiel ou synthétique? Les fruits sont des mines de molécules utiles et encore non exploitées: Le POPj est un nouveau colorant jaune extrait des pommes très intéressant." *Pour la science* 348 (October 2006): 4.

SIX

ARTISTIC CHOICE + CULINARY NOMENCLATURE

BEFORE WE TAKE A CLOSER LOOK at the fears and uncertainties that the adoption of a radically new approach to cooking is bound to arouse, let me continue to try to make the strongest possible case on behalf of note-by-note cooking, which in any case is rapidly being brought into existence in many parts of the world. Shapes, consistencies, tastes, odors, trigeminal sensations, and colors can now be created independently of one another, or very nearly so. The main problem facing us at this point is that there are too many possibilities!

The chef is seated before his piano. What kind of culinary music will he decide to play? For a traditional cook, the answer is quite simple and every bit as boring. In France in the springtime, cooks living in the countryside—in the Jura, for example—who go into the forest to pick the first morels will prepare them in a cream and yellow wine sauce to accompany a roasted chicken; cooks living in cities, in the Jura and elsewhere, who buy the season's first asparagus in their local market will serve them with a sauce chosen from the classical repertoire, which in many cases is restricted by rather peculiar conventions. Cooks in Brittany will likewise work with whatever is in season: scallops, perhaps, or turtle or oysters. And so it will be in every other part of the country.

The seasons, the land, regional tradition—thus the three pillars of traditional cooking, each one perfectly legitimate in its way. In combination, however, they amount to a sort of unspoken agreement that no one will have to think very hard about what to cook. Still today the most inventive chefs, even the least hidebound among them, go on being parties to it whenever they content themselves with devising more or less minor variations on a set of age-old themes. Note-by-note cooking threatens to topple this whole system. No compound known to chemistry is uniquely associated with a particular place. No compound has a particular season. As for conventions, there aren't any yet. Everything remains to be invented.

SUBSTANCE AND FORM

Note-by-note cooking allows us to artificially recreate traditional dishes, to make artificial wines (which should not be called "wines," as we have already seen), artificial cheeses (some have already been made), and artificial fruits (raspberries, for example). None of this is at all difficult from the technical point of view. But is there really any point to it? The fact remains that a copy is not exactly the same as the original, and there is a risk that it will be criticized (wrongly, in my view) for just this reason. Inevitably, the old and the new will be compared.

Let's approach the question from another angle, resorting once again to an analogy with music. Just as electronic synthesizers can be programmed to play "Twinkle Twinkle Little Star" with specific and novel timbres that nevertheless are neither strident nor loud, so one might create a comfortable, familiar, unchallenging version of note-by-note cuisine. Or one might create a new world of provocative, even abrasive flavors. There is also an analogy with painting. Painters, using modern pigments, can perfectly well make traditional figurative works, but they can also make works so strange that viewers who have not made the effort to learn a new pictorial language cannot help but angrily denounce them. "That's not art!" they cry.

Earlier, in chapter 2, we saw how useful tables can be in stimulating the imagination. Here is another one (table 6.1), for chefs brave enough to position themselves in the emerging market for culinary art. Cooks are condemned

	OLD FORM	NEW FORM
Old Substance	In music, a composition such as "Twinkle Twinkle Little Star" may be likened to a natural substance. The flute gives it an ancient, traditional form. In other words, "Twinkle Twinkle Little Star," played on the flute, is old music played in an old way. In painting, figurative representation, in which the things we see in the world around us (trees, persons, and so on) are realistically rendered, may likewise be thought of as depicting old substances. By applying the same pigments that the painters of the Renaissance used (just as the flute can be used to make music still today), we can give traditional scenes an equally traditional form. Think of the marvelous *Isenheim Altar* in the Unterlinden Museum in Colmar. And in cooking? A dish of choucroute, made with cabbage, sausages, and potatoes, is an old substance in an old form. With note-by-note cooking, it is possible to give choucroute a perfectly traditional interpretation, however uninteresting it may be to do this.	Let us take an old substance and give it a new form. It is the most daring of the four possibilities summarized here. "Twinkle Twinkle Little Star" can be played on a synthesizer. If the synthesizer is used to produce flute sounds, the work's new appearance will be imperceptible, or perceptible only by highly trained listeners. By contrast, if a synthesizer is used to produce new sounds, unknown to classical music, "Twinkle Twinkle Little Star" will be perceived in a new way, and the hold of the old form over us will be weakened. Similarly, in painting, no traditional painter would have drawn three angels as Salvador Dalí did. With Dalí, there was an old substance (the angels) and a new form. So, too, though in a rather different spirit, the use of acrylic paint by Andy Warhol and others gave traditional scenes a new look. In cooking, a choucroute composed note by note would fall under this category so long as its appearance is novel. Molecular cooks have already been doing this sort of "deconstruction" for a number of years now. It has now become possible to play the same game at a higher, more challenging level.

(CONTINUED)

OLD FORM	NEW FORM
New Substance	Here a new substance is cast in an old form. In music, I think particularly of the work of Iannis Xenakis, who combined human voices and traditional instrumentation to create utterly novel compositions. Later in this chapter, I will briefly consider what inspired him to do this.	Giving a new substance a new form is the most daring of the four possibilities summarized here. At the beginning of the twentieth century, with the advent of electroacoustical instruments, musicians such as Pierre Schaeffer, Edgard Varèse, and Karlheinz Stockhausen were able to make novel sounds, which could be used to compose music in both classical and modern styles. The new melodies and tones introduced in some of their works sounded like screeching cats to lovers of traditional music. But apart from the novelty of the form, the substance of these works is so unfamiliar that our minds have not yet grown fully accustomed to them even today.

New Substance (row continued):

OLD FORM:

In painting, many modern artists have created new substances with traditional pigments—Zao Wou-ki and Pierre Soulages, for example.

In cooking, it is likewise a matter of using traditional ingredients to obtain new effects. Pierre Gagnaire, for example, is able to do quite extraordinary things with quite ordinary ingredients: Camembert with raspberry sauce, tuna and veal tartars, a baba with button mushrooms (rather than rum), scallops with licorice sauce, and so on.

NEW FORM:

In painting, too, artists such as Yayoi Kusama make works that are unprecedented with regard to both form and substance.

In cooking, I invite those who accept the challenge of note-by-note cooking to throw themselves into it. We can be certain that their efforts will be praised by the moderns among us and reviled by the ancients, just as the innovations of the first generation of molecular cooks were. Let us encourage the most forward looking of the new pioneers: they and their followers will be the ones left standing at the end of the day.

to be either artisans or artists; if they are artists, they can practice their art only if they have an audience—a public.

This table will naturally provoke all kinds of reactions, but everyone will agree that the question at issue here is a very difficult one. I shall begin by quoting from an unpublished lecture given by Iannis Xenakis in 1983 to the Atomic Energy Commission (as it was then known) in Paris, "Current Problems of Musical Composition" (followed by dinner at La Tour d'Argent—proof, one is tempted to say, that the prospect of fine food is an indispensible condition of originality):

> In 1953, I proposed a new conception of music. It rests on the mass effect of individual sound events, an effect that prevents individual events from being followed separately. As in the song of cicadas in summer, the noise of crashing waves or of a hail storm, the slogans shouted during demonstrations by tens of thousands of men and women, especially during bloody clashes with the authorities. Thus, instead of constructing on the basis of melodic lines or series of notes, on the basis of distinct sequences of sound events, it means creating architectures with masses of events and modulating these masses as a sculptor would, only using tonal material. To do this, I needed to use a probability calculus. Hence stochastic—probabilistic—music.
>
> An orchestra composed of a hundred musicians has offered until now, and always will offer in the future, remarkable opportunities for exploration. If each musician can play, on average, five notes per second, that comes to five hundred sound events per second in all, well beyond our mental ability to tell them apart. It is here that our brain intervenes. In order to overcome a statistical impossibility, it invents concepts of average density, uniform distribution, degree of order, rate of transformation, degree of agitation, and so on. There is a parallel here with the kinetic theory of gases in the nineteenth century, which later became statistical mechanics. I have composed works such as *Metastasis* and *Pithoprakta* in which Poisson and Gauss distributions—linear, uniform, and so on—are combined in an abstract black box that is the architecture of this new music.

Some people like Xenakis's music, others find it unlistenable. Many people have trouble with modern art in general, not just modern music or painting. In cooking, it sometimes happens that we find combinations of perfectly traditional ingredients shocking because the combinations are unfamiliar to us. Culinary artists who grow weary of doing again and again what has already been done thousands of times before can take heart in knowing that works of all kinds, even ones that we find jarring at first, will be accepted in the end—on one condition, that they do not lose us completely. But the converse is also true: novelty is not in and of itself a guarantee of artistic worth. In literature, formal innovation is not solely responsible for the effects achieved by the greatest modern authors; indeed, taken too far, formalism is a recipe for tedium. The reason mere novelty is not enough by itself is that there must be an inner resonance, a deeply felt emotion, that unites the reader with the spirit and purpose of the work.

In music, there is melody and rhythm; in painting, shape and color. In either case there is structure and contrast: structure, because an unordered series of sounds or images strikes us as incoherent; contrast, because juxtaposition and opposition are what we perceive when we see or listen to a work of art. Likewise, in cooking, contrast is essential because our sensory system is constructed to detect it; structure is essential as well because contrast cannot exist without it. These two fundamental concepts may be analyzed under the categories of shape, consistency, taste, odor, trigeminal sensation, and color, as we have already done.

Finally, the note-by-note cook will have to ask himself if it is worth going to the trouble of making new dishes that are rather timid and restrained out of a desire not to offend more traditionally minded food lovers, or whether he is prepared to do what is necessary to shock settled sensibilities by creating dishes that are daring in both their spirit and their form. Connoisseurs of powerful sensations will not complain if he chooses the latter course.

THE CONSTRUCTION OF FLAVORS

The question of what we should put on our empty plate has not yet really been answered. Intelligent and resourceful cooks will respond in a practical fashion, by taking as their point of departure basic, elementary things—which is to say whatever compounds they can lay their hands on. Just as a fine turbot may inspire traditional and modern cooks alike to produce original works of art, some compounds naturally suggest new paths to explore. Limonene, with its fresh lemony scent; cinnamaldehyde (cinnamic aldehyde), which is hard to tell apart from cinnamon; octenol, with its odor of mushrooms and forest undergrowth—these and all the other compounds we have looked at are so many invitations to culinary creativity.

A few years ago, collaborating on a menu for a dinner of molecular cooking, the pastry chef Nicolas Bernardé and I had the idea of trying to work from a pure feeling. After all, the work of a truly talented chef is animated by personal emotion—not culinary emotion, mind you (for that would almost unavoidably lead him to reproduce something very classic), but instead a feeling such as the one he experienced as a child, for example, going out to pick mushrooms for the first time with his grandfather. At the edge of the forest, the boy comes upon a magical scene: the tree trunks bathed in dark shadow, the lush green of the moss, the rays of light filtering through the branches. It was just this emotion that Bernardé wished to convey, as it happens, and during the meal that followed I saw men and women with tears of joy in their eyes: so well had the chef succeeded in sharing this emotion, it brought back to each one of them his or her own childhood memories of the countryside.

Once the guiding idea has been decided upon, the dish must be thought out in every detail before it can be brought into being. The dish will need to embody a whole set of specific organoleptic characteristics (consistency, odors, tastes, and so on). In theory, the preceding chapters will have told us everything we need to know in order to convert inspiration into reality. As a practical matter, however, a great deal of experiment and further reflection will be needed.

Though the mode of culinary deconstruction promoted by molecular cooking may be justified in various ways, it seems to me to have no value whatever by comparison with the crucial imperative of construction. Note-by-note cooking is meant not least of all to remind us that reproducing familiar dishes is lazy and unworthy of the serious cook. At bottom, I believe, cooking must concern itself with constructing new dishes, not only from the point of view of shape and consistency, but also of taste and smell and all the other sensations that contribute to the experience of eating. Is there really any difference between construction in this sense and Carême's conception of architecture? Not at all—and yet the idea of culinary constructivism is not something that would have occurred to anyone two hundred years ago. One way of appreciating its originality is to think of flavor as a sensation that is indissociable from the phenomenon of duration. A comparison with language may be illuminating. To adopt the linguist's terms, one may say that the act of eating is organized along a syntagmatic axis and a paradigmatic axis: at any given moment we experience a variety of sensations, each of which changes in the course of mastication, even afterward in some cases. A skillful note-by-note chef will have organized all these sensations in advance, constructing them in the first instance and then over all those instants that follow—which is to say over time.

The enlightened cook will begin by reasoning from simple physiochemical facts. From the fact, for example, that salt reinforces the sensation of sweetness and weakens the sensation of bitterness, or that strawberry and orange-blossom water jointly produce a flavor of wild strawberry. Imagining how different ingredients will combine with one another is a matter of knowing beforehand which ones are robust enough to withstand the effect of the others, how the individual tastes will blend together, how a change in our perception of one or more parts will influence our perception of the whole.

Next, the cook must come to terms with the phenomenon of duration, the ebb and flow of sensations over time. Depending on the physical composition of a dish—the ingredients themselves, the presence of fatty substances, the colloidal structure (foam, emulsion, gel, suspension), and so on—various flavors appear and disappear at various moments in the course of tasting. For example, the longer a mayonnaise (the prototype emulsion for our purposes) is beaten, the more the odorant components will stand out, whereas in a mayonnaise that has been beaten for a shorter time the sapid components will be more prominent. Or again: fatty substances coat the mouth and lengthen the time during which odorant compounds are perceived. And again: different molecular combinations modify the relative intensity or assertiveness of different flavors. Cooks already have an intuitive sense of what this new kind of constructivism entails. When they decorate a sauce with a sprig of chervil, for example, they introduce an element that needs to be chewed, lengthening the period of time over which the flavor of the sauce is perceived. There is much, much more to be said in this connection, but rather than repeat myself here, I must thank my readers for their patience in waiting until my next book, *Explorons la cuisine*, appears in English.

For the moment, it is enough that we frankly acknowledge how very difficult the culinary art is. Cooking at a high level, whether in traditional or note-by-note fashion, demands a great deal of technique and an even greater amount of feeling. Technique can be learned. But what about feeling?

NAMING DISHES

Last, but hardly least, there is the question of what we are to call note-by-note dishes once we have created them. In April 2009, Pierre Gagnaire named the first dish in the history of note-by-note cooking "Note-by-Note No. 1." One could carry on in this vein (No. 2, No. 3, No. 4, and so on), of course, but that would seem rather cold and impersonal, don't you agree? For the names we give to dishes are very much a part of why we like them. Think of a choucroute royale, whose name aptly gives the impression of something more lavishly appointed and more copious than a simple choucroute. Let's make a brief detour through the past, then, in the hope of gleaning a few ideas from those who came before us.

BEFORE CARÊME

In the beginning, when human beings had only recently parted ways with their primate ancestors, when language was still rudimentary, they ate without giving names to what they ate. In this respect, at least, they were no different than animals. Humans eventually acquired the faculty of speech and with it the habit of distinguishing between different foods, initially by means of a simple phrase that identified the food itself and its method of preparation. We do the like of this still today when we speak of roasted chicken, grilled steak, fried potatoes, sliced tomatoes, hard-boiled eggs, and so on. The famous cookbook by Guillaume Tirel (ca. 1310–1395, better known as Taillevent), *Le Viandier,* which first appeared in print in the late fifteenth century, carried on an already ancient tradition: boiled beef ("bouture de grosse char"), boiled venison ("chevreau sauvaige"), boiled wild boar ("sanglier frais"), baked chicken with cumin sauce ("comminée de poulaille"), and so on.

Once the ingredients had grown beyond a certain number, or when a particular cooking procedure came to be considered of greater interest than the food itself, new names were introduced: pot-au-feu, vinaigrette, white sauce, stew of small birds, and so on. Thus one finds preparations named after their most salient ingredient (a vinaigrette, for example, is notable especially for the vinegar it contains) as well as mention of the cooking equipment used to make certain dishes (a pot-au-feu is made in a pot over a fire); in some cases, the distinctive property of a preparation is emphasized (a white sauce, after all, must be white).

Soon thereafter the names of dishes came to denote their place of origin: Taillevent's *Viandier* speaks, for example, of a sauce from Poitou ("saulce poitevine"). *Le Ménagier de Paris,* a book by an anonymous author published just before Taillevent's death, speaks not only of savory pastries from Italy ("tourtes pisaines" and "tourtes lombardes") but also of a Savoy broth ("brouet de Savoie") and a German broth with eggs poached in oil ("brouet d'Alemaigne d'oeufs pochés en huile"). But these names are still less common than references to the manner in which a dish is prepared or served: capon pie ("pastés de chapons"), twice-cooked pike and eels ("bécuit de brochets et d'anguilles"), whitebait with cold sage sauce ("ables et froidc sauge"), venison with wheat boiled in milk with sugar and spices ("venoison à la froumentée"), roast Provencal figs covered with bay leaves ("grosses figues de Prouvence rosties et fueilles de lorier par-dessus"), cooked apples ("pommes cuites"), and so on.

The tendency to identify dishes with the meats, fish, and vegetables of which they were chiefly composed was reinforced by the appearance of the French translation of a book by the fifteenth-century Italian humanist Bartolomeo Sacchi (who wrote about cooking under the pseudonym "il Platina"), *Le Platine en françois* (1505). The section titles of Sacchi's book correspond to more or less general categories: "Venison," "Cabbages," "Brown Meagre" ("corbeau de mer," a kind of fish), "Gourds," and so on. One does, of course, encounter dishes such as cabbage à la romaine, squash à la cathalane, and cold eggs à la florentine, but these culinary genealogies were no less anecdotal than they had been a century or two earlier. Names now began to be attached to specific preparations, particularly in the case of soups: one, made from elder flowers and known as zanzarella ("potaige des fleurs du seuz appele zanzarelle blanche"); another, a Lenten broth called leucophage ("potaige en jeusne appele leucophage"); or again, a soup with pasta made from wheat and cut into the thin pieces from which it took its name ("potaige frumentin ou menudets").

A century and a half later François Pierre de la Varenne's book *Le Cuisinier françois* (1651) makes it clear that naming conventions were now more firmly anchored in technique, which itself had been refined. La Varenne uses terms such as *estuvé, court-bouillon, ragoust, fricasée, baignets, bisque*—all of them familiar still today despite minor changes in spelling. But some are new. He speaks of pike, for example, that has been cooked for only a very short time in order to bring out the blue hues of its skin ("brochet au bleu"). And when he mentions a dish such as champignons à la crème, the name no longer tells us everything we need to know. It is not a matter simply of cooking mushrooms in cream: the name, though it seems to describe the ingredients, as in older cookbooks, is actually a very brief summary of a rather elaborate recipe. Above all, one is struck by the multiplication of phrases indicating a style of preparation associated with a particular person, place, or nation: "Pièce de bœuf à l'Angloise" (or "à la Chalonnaise"); "Potage à la princesse" (or "à la reyne"); "Oeufs à la portugaise" (or "à la Varenne"); "Pasté à la Cardinale" (or "à l'anglaise"), and so on.

In *L'Art de bien traiter* (1674) by L. S. R. (an anonymous cook whose initials may stand for "le sieur Robert"), two types of name are frequently met with. In addition to fricasées of pig's feet and marinated chickens served with a sauce, one finds suckling pig in the style of Père Douillet and stuffed legs of mutton à la royale. L. S. R. criticizes La Varenne for the needless complication of his recipes, calling them "inutile" and "dégoutant" (the latter term still bearing its original connotation of excess rather than the modern sense of something that is sickening or loathsome)—though his own cooking is rather complicated, enough so at least that it has now become quite impossible to discover how to make a dish from its name alone. One now begins to see much longer names, such as "Échinée aux pois et petits oisons à la daube mangés chauds, et fricandeaux de veau frits en beignets ou en ragoût ou piqués rôtis." In this dish, for example, one begins by blanching the meat, which is then pounded and cooked in a pot with a great many ingredients, including lard, salt and spices, fines herbes, broth, onions studded with cloves, streaked bacon, lemon, white wine, and artichoke bottoms.

Let's take another great leap forward. Almost two centuries later, in 1847, the final volume of Marie-Antoine Carême's monumental *L'Art de la cuisine française au XIXe siècle* was posthumously published. With Carême, a distinction is made for the first time between basic preparations (bouillon, mirepoix, meat and fish fillings for various kinds of dumplings) and the recipes themselves. Newly baptized dishes predominate: "Potage de santé à la française" (or "à la régence," "au chasseur," etc.), "Garni de poulet à la reine" (or "à la paysanne," "à la Crécy," "à la Buffon," "à la Girodet," "à la Monglas," "à la princesse," "à la Rossini," etc.). It hardly comes as a surprise that a queen or a princess who particularly liked a certain dish was able to arrange for it to be named after her. One's attention is drawn instead to a pattern of growing complexity in the culinary arts, which now distinguish between local, regional, and national styles of cooking while at the same time indicating levels of social stratification. Moreover, advances in communication gave much wider currency to names than in the past, and this at a time when popular interest in cooking was growing very rapidly (to judge from the increasing demand for cookbooks). Carême's student Urbain Dubois ably upheld and extended the tradition established by his teacher: "Purée de topinambours Palestine," "Purée de choux-fleurs Dubarry," "Purée de concombres Mathilde," "Purée d'asperges comtesse," "Purée de haricots frais Musard," "Purée de pomme de terre Jackson," "Gros merlan à l'intendant," "Moyens merlans à la diplomate," "Rougets à la Colbert," and so on.

ESCOFFIER AND AFTER

By the beginning of the twentieth century, marked in the world of cooking by the appearance of Georges-Auguste Escoffier's *Le Guide culinaire* in 1903, names had not gotten any shorter: "Coquille de queues d'écrivisses Cardinal," "Filet de veau Agnès Sorel," "Poularde pochée à l'Anglaise," "Suprême de volaille Valençay," "Côtelettes de pigeonneaux en chaud-froid." By this point, however, so miserably did the name of a dish fail to describe its component parts that cooks now began to feel a need for a guide to *Le Guide*, something that would help them recall the various ingredients used to make the thousands of dishes that the master had codified. Some twenty-five years later this aide-mémoire conveniently materialized in the form of a book by Escoffier's associates Thomas Gringoire and Louis Saulnier, *Le Répertoire de la cuisine* (1929).

The weakness for pompous and often precious names was further aggravated by the alliance of cooking with an overtly literary sensibility. In *Éloges de la cuisine française* (1933), which I mentioned earlier, Édouard Nignon carried the practice of using self-congratulatory epithets to describe dishes to new lengths (everything is "mignon," "superbe," "succulent," "frais," "tendre"), while never succumbing to the temptation of choosing a simple name when a more complicated one could be devised: "Tartelettes de mauviette à l'infante," "Porcelets à la brioche à la française," "Paupline de jambon Valonde," "Grenadins de gélinottes à la Bariatinski," "Escalope de langouste régence," "Crème à la freux aux amandes fraîches," "Grenadins de barbues à la Chartres," "Caprices de dame fleurise," "Pascaline de fois gras à la Rohan," and so on. Still, it must be conceded that Nignon's cooking was remarkably good, and in any case well deserving of the praise it heaped upon itself.

Skipping over a few more decades, we come to Paul Bocuse and *La cuisine du marché* (1976). Here at last we find a more sober and restrained style, with "Abricots Colbert," "Agneaux de pré-salé," "Aloyaux rôtis," "Beignets d'anchois," "Choux verts," "Coqs au fleurie," "Crêpes à l'eau de fleur d'oranger," and so on. While Bocuse's cookbook does not have the same ambition as either of the two treatises I have just mentioned, by Nignon and Escoffier, its recipes are nonetheless worked out with considerable sophistication and attention to detail. Making a cucumber salad, for example, is not simply a matter of slicing the cucumber and adding a vinaigrette sauce to it, as nearly every novice would do. No, the cucumber must be peeled and cut lengthwise in two; then, once it has been seeded, you have to cut each half into very thin slices, arrange them on a plate, and sprinkle them with salt, then mix them together in a bowl and let them macerate for an hour until all the excess water has been drawn out, then carefully strain and season with pepper, oil, vinegar, and chopped chervil.

It is only fitting that I leave the last word to Pierre Gagnaire. On a 1977 menu, one finds a dish called "Pochette du saint-pierre et des poivrons." The name does not tell you that in addition to the John Dory and bell peppers there are also onions, cucumbers, lemon, and white wine. A few years later the number of ingredients mentioned begins to rise. In 1983, for example, one finds a "Gelée d'huîtres et foie gras de canard au jus de betterave, tartine de seigle au beaufort"; in 1988, a "Cuillère de gelée de boeuf à l'huître, sablé aux courgettes, served with a Moussette de lapereau au genièvre," is followed by "Noix de Saint-Jacques rôties au poivre de Sichuan, navets confits au foie gras et dominos de poires au beurre frais," garnished by a "Cuillère de gelée de boeuf à l'huître." As the preparation becomes more involved, the name grows longer as well. From 1990: "Socca de morue au vieux parmesan, chair de tourteau, feuilles de choux de Bruxelles et noix de Saint-Jacques à la croque-au-sel," napped with a "Jus d'oursin au vin jaune du Jura." The names are not always so long, however. From 1991: "Turbot au café, petits oignons à la cardamome," with a "Jus de tomate à l'aurore, tomate fripée." But sometimes they are even longer—and more emphatic. Come 2005, one finds this bell ringer: "Le Noire: Crémeux de riz NOIR vénéré au poivre NOIR de Sarawak; radis NOIRS aux quetsches," accompanied by a "Belle noix de ris de veau braisée dans un jus d'olives NOIRES de Nyons aux trompettes" and a "Gelée de navet demi-deuil et vent des sables aux pitchounes."

Let's be honest, it hardly matters what these dishes are called—in French or in English. I have left their names untranslated because they are not meant to be attempted by home cooks, who would have to start work a week in advance, if not several weeks—even assuming they could obtain all of the ingredients called for, some of them exceedingly rare and expensive.

THE FIRST GENERATION OF NOTE-BY-NOTE MENUS

Note-by-note cooks are perfectly free, if they like, to recapitulate the history of cooking in selecting names for their creations. One might begin very simply, perhaps by naming dishes after their chief structural characteristic:

"Brittle Polyphenol Gel," "Beet-Powder Foam," and so on. One could also give them names associated with historical periods or literary figures: "Citric Acid Conglomèles à la Renaissance," for example, or "Fibrés à la Balzac." Some may wish to emphasize that the new movement carries on an older tradition of cooking: "Soufflé of Whole Syrah Triglycerides and Polyphenols"; "Whisked Emulsion of Glucose, Glycine, Proline, and Hydroxyproline." Others may prefer to call attention to their inventions by keeping things simple.

Myself, I expect to see anything and everything. When it comes to names, we have already seen it all in cooking—and there is no reason to think that the future will be any different. The artist is a singular person, who by his very nature cannot help but try to set himself apart. Here are some of the wondrously strange names that have already been given to the earliest note-by-note dishes.

DISHES BY PIERRE GAGNAIRE

Note-by-Note No. 1 (April 24, 2009)

Apple Pearls, Opaline, and Lemon Granita (April 2009)

A Savory Pastry (May 2009)

Mace-Seasoned Artichoke Velouté, Raw Provençal Artichoke Bouquet, and Yellow-Wine Aspic Squares (2006)

Polyphenol Caramel Disks (2009)

ÉCOLE LE CORDON BLEU DINNER (PARIS, OCTOBER 16, 2010)

Royale of Undergrowth, Truffled Blancmange, and Slightly Foamy Bouillon

Iodized Octopus and John Dory Flesh, Spaghetti Foam, and Transparence with Cepe Mushrooms

Twice-Cooked Young Pigeon with Stewed Thigh Meat, Winter Squash (*Potimarron*) Fondant, Polyphenol Gel, Virtual Asparagus

Fresh Goat Cheese Chantilly

Napoleon Marshmallows, two textures

POTEL & CHABOT DINNER (PARIS, JANUARY 26, 2011)

Oyster Tapioca Amylopectin Foam Bavarois, Lemon Tapioca

Sea Water Jelly, Cream of Oyster Soup, Wind Crystals

Lobster Soufflé, Wöhler Sauce, Raspberry Agar Jelly

Beef and Carrot "Fibré," Cappelini, Turned Carrots

Beef Cheek with Brown Gravy

Cassis Powder Eruption, Cassis Boule

📖 ÉCOLE LE CORDON BLEU DINNER (PARIS, OCTOBER 15, 2011)

Note-by-Yolk (in the style of an oeuf en meurette)

Tricolored Surf-and-Turf Napoleon with Kientzheim and Shellfish Sauce Duo

Note-by-Note Version of a Pot-au-feu

Mozzarella Reconstructed, with olive oil and lamb's lettuce (*mâche*)

Cordon Bleu Dessert

And the future has only just begun!

FURTHER READING

Bocuse, Paul. *La cuisine du marché: En hommage à Alfred Guérot*. Paris: Flammarion, 1976. Available in English as *Paul Bocuse's French Cooking*, translated by Colette Rossant (New York: Pantheon, 1977).

Carême, Marie-Antoine. *L'Art de la cuisine française au dix-neuvième siècle: Traité élémentaire et pratique*. 5 vols. Paris: J. Renouard et Cie., 1833–47; reprint, Paris: de Kerangué & Pollés Libraires-Éditeurs, 1981.

Escoffier, Georges-Auguste. *Le Guide culinaire*. 4th ed. Paris: Flammarion, 1921. Many times reprinted by the same publisher; the English translation by H. L. Cracknell and R. J. Kaufmann (New York: Mayflower Books, 1979) has likewise been regularly reissued, most recently by Wiley.

Gringoire, Thomas, and Louis Saulnier. *Le Répertoire de la cuisine*. Paris: Dupont et Malgat, 1929. This book remains in print still today with Flammarion, though again it seems never to have been translated into English.

Le Ménagier de Paris. Translated into modern French from the 1393 edition by Georgina E. Brereton and Janet M. Ferrier. Oxford: Clarendon Press, 1981. Available in English as *The Good Wife's Guide: A Medieval Household Book*, translated by Gina L. Greco and Christine M. Rose (Ithaca, N.Y.: Cornell University Press, 2009).

Nignon, Édouard. *Éloges de la cuisine française*. Paris: H. Piazza, 1933. Reprinted most recently by Flammarion in 2009. No English translation exists.

L. S. R. *L'Art de bien traiter*. Reprinted from the 1674 edition together with Pierre de Lune, *Le Cuisinier* (1656), and Audiger, *La Maison réglée* (1692), in Gilles Laurendon and Laurence Laurendon, eds., *L'art de la cuisine française au XVIIe siècle* (Paris: Payot & Rivages, 1995). As in the case of Carême's work, no English translation has yet been made.

Robbe-Grillet, Alain. *For a New Novel: Essays on Fiction*. Translated by Richard Howard. Evanston, Ill.: Northwestern University Press, 1965.

Sacchi, Bartolomeo (Battista Platina). *Le Platine en françois: D'après l'édition de 1505*. Edited by Silvano Serventi and Jean-Louis Flandrin. Houilles, France: Manucius, 2003.

This, Hervé. *Building a Meal: From Molecular Gastronomy to Culinary Constructivism*. Translated by M. B. DeBevoise. New York: Columbia University Press, 2009. See especially the discussion of culinary constructivism in the last chapter, 109–121.

This, Hervé, and Pierre Gagnaire. *Cooking: A Quintessential Art*. Translated by M. B. DeBevoise. Berkeley: University of California Press, 2008. See especially the chapter "The Present and Future of Cooking," 259–326.

Tirel (Taillevent), Guillaume. *Le Viandier d'après l'édition de 1486*. Edited by Mary Hyman and Philip Hyman. Houilles, France: Manucius, 2003. See the English translation by James Prescott, *Le Viandier de Taillevent*, 2nd ed. (Eugene, Ore.: Alfarhaugr Publishing Society, 1989), also available in an online version at http://www.telusplanet.net/public/prescotj/data/viandier/viandier1.html.

Varenne, François Pierre de la. *Le Cuisinier françois*. A facsimile of the 1651 edition with a preface by Mary Hyman and Philip Hyman. Houilles, France: Manucius, 2003. A modern English translation with commentary by Terence Scully has recently appeared as part of an omnibus volume, *La Varenne's Cookery: The French Cook; The French Pastry Chef; The French Confectioner* (Totnes, U.K.: Prospect Books, 2006).

NUTRITION, TOXICOLOGY, MARKET DYNAMICS, PUBLIC INTEREST

I SHOULD LIKE TO CONCLUDE by briefly taking up a series of topics that will need to be considered more carefully if note-by-note cooking is to win broad popular acceptance: the nutritional value of pure compounds; the level of toxicological risk entailed by their use; the selection of suitable compounds, and their commercial availability; and the political issues raised by note-by-note cooking in relation to agricultural production, food safety, energy use, and public education.

THE MIXED BLESSINGS OF ABUNDANCE

Humanity is at a turning point in its alimentary history. In the industrialized countries, our generation is the first not to have known famine. At the same time we are witnessing the early stages of an obesity pandemic because our bodies were not designed to cope with the abundance of foods presently available to us and because older styles of cooking, which developed in a world of far more limited resources than today, are no longer appropriate to the situation in which we find ourselves. Fortunately, our brain is an adaptive mechanism that can help us find ways of overcoming this crisis—if only we put our minds to it! Note-by-note cooking is a product of human ingenuity that will both concentrate our attention and reward the application of intelligence and imagination.

I must begin, however, by confessing my own incompetence in this matter, and perhaps that of nutritionists as well. It is, of course, well known that foods must provide the human organism with water (evaporated chiefly through the skin) and enough fuel to produce the energy we need to live every day, on the one hand, and to replenish the body's energy reserves as they are depleted, on the other. Food in its traditional form therefore consists of three primary substances: lipids, proteins, and carbohydrates. Quantity isn't everything, however. In order to function, we need particular kinds of lipids, particular kinds of proteins, and particular kinds of carbohydrates—all combined in the right proportions.

First, lipids. All cells are bounded by a membrane, which consists of a double layer of phospholipids. Our bodies manufacture these compounds from the fatty substances we ingest. They contain mainly triglycerides, which we discussed earlier. A triglyceride is an organic compound with three fatty-acid residues bonded to a glycerol molecule. Digestion releases these and other fatty acids, which then pass into the bloodstream. (Since we are constantly being bombarded with dietetic information these days, perhaps I need not add that there are different sorts of fatty acids, and that the molecular properties of various triglycerides depend on the precise proportions in which their constituent compounds are combined. For the moment, however, these proportions are not fully known.)

Second, proteins. Their digestion releases peptides and amino acids. These must be well suited to the biochemical needs of the human organism: they must be "balanced," as the nutritionists are fond of saying. Eating meat and fish provides us with a combination of proteins similar to the ones that make up our own flesh, but by themselves they do not constitute a healthy diet. Many other animals eat insect larvae, for example, but the unbalanced amino acid content of such meals must be supplemented by other kinds of compounds.

Carbohydrates form the third element of the triad. Earlier we saw that glucose, slowly released by modified starches, for example, plays an important role in human energy metabolism. Another carbohydrate, cellulose (which, like glucose, is a polysaccharide, though an indigestible one), is used as a bulking agent in various processed foods. Note-by-note cooks may find it useful to incorporate cellulose fibers in preparations that will be regularly consumed over an extended period.

These three groups of compounds, essential though they may be, are not the only things we need in our diet. Medical research on scurvy, a disease to which sailors making long voyages at sea were particularly prone in the past, led finally in the early twentieth century to the discovery of vitamins, another group of compounds that are indispensable for vital functions. Similarly, the study of cretins of the Alps, as they were called—mountain peoples in Europe suffering from goiter, an enlargement of the thyroid gland caused by a lack of dietary iodine—led to the discovery of oligoelements satisfying specific nutritional requirements. Today oligoelements are reasonably well understood, with regard to both their chemical properties and the quantities in which they must be consumed in order to sustain life.

Nutritionists have their work cut out for them. An enormous amount of research will be required if we are to be able to construct note-by-note dishes having the same nutritive value as the foods we have grown up eating. This will be a labor of decades. But the time and effort will have been well spent, for the challenge of feeding as many as nine billion human beings by the end of the twenty-first century promises to make the replacement of traditional cooking by note-by-note cooking unavoidable. For the moment, however, no one should fear that choucroute, coq au vin, and all the other delicious items of the traditional repertoire (delicious, at least, when they are well prepared) are endangered species. We should rejoice instead at the prospect of a multitude of splendid new creations being added to them. Even if note-by-note dishes are not nutritionally balanced to begin with, their dietary value will steadily increase as scientific research gathers pace. In the meantime, they will offer us marvelous and utterly novel sensations.

I invite anyone who asks whether the search for novelty in cooking is a futile endeavor to consider music. Is it futile that music, which has delighted human beings for thousands of years, should forever go on developing in new ways? Is it futile that painting goes on changing? Sculpture? Literature?

A WORLD OF PLENTY, FILLED WITH DANGER

Earlier I approvingly quoted the biblical reference (in Deuteronomy) to the good Lord's rich storehouse. Yet we should not be so naive as Voltaire's Pangloss and Candide, who considered our world to be the best of all possible worlds. Our world is full of dangers, and our humanity—what makes us human beings and not merely animals—is but a long apprenticeship to prudence, which as creatures of a hostile planet we neglect only at our peril.

I do not mean that nature does not provide us with what we need to survive. If it didn't, we wouldn't be here to talk about it today! But nature is not therefore our friend. Anyone who eats a wild vegetable without knowing exactly what it is risks being sickened, even killed by it. In this, as in every other aspect of our lives, we are constantly obliged to take precautions. The human race has long been incapable of living naked, having lost its protective coat. We need clothes (which are artificial, of course, because we have made them ourselves) for protection not only from the sun, but also from the rain and the cold. Without medicines, without hygienic products and certain kinds of cosmetics, we would not live far past reproductive age. It is only on account of human contrivance and invention that average life expectancy in a few fortunate countries of the world has now reached eighty years or more. Nothing could be more plain: nature is not always good; if it were, humanity would not have to shield itself against it at every turn.

In earlier chapters I touched briefly on the dangers associated with the use of compounds in note-by-note cooking. Quite obviously the reason for concern is very real. But the way in which the question is usually posed—Is this or that compound good for you?—is too general to serve any useful purpose. Every time you hear the phrase "good for you," pay attention to the context in which it occurs. A fatty acid that is good for the heart may nonetheless happen to be bad for the brain. A certain antioxidant phenolic compound may protect against heart disease, but it may nonetheless cause other pathologies. Conversely, an alkaloid that is popularly thought to be injurious to health may turn out to have benign pharmaceutical applications that become known only as a result of rigorous testing.

The problem is therefore by no means a simple one. Paracelsus, in insisting on the importance of administering the right dose, laid down a sensible rule, it seems to me, for it is often true that a compound can have harmful effects only when it is given in amounts substantial enough to make it probable that a large share of its molecules will reach their receptors. This is why toxicologists consider a range of statistically likely outcomes. Toxicity usually appears only beyond a certain level of concentration, but in some cases it may be manifested immediately, and in others it may happen that low doses are beneficial, whereas higher doses have harmful effects. The vegetable kingdom is so varied, however, that in an even greater number of cases we know nothing, or next to nothing, about the toxicological effects of its members. Even so, scientific knowledge in this domain is not inconsiderable. It would be a crime not to put to good use the results patiently accumulated over many years by pharmaceutical researchers, who with the advent of note-by-note cooking will have a new scientific mission, and indeed an entirely new social role, that goes well beyond investigating the chemical properties of poisonous mushrooms.

Here again the Internet will perform a valuable service by allowing note-by-note cooks to quickly locate information about the known dangers posed by various compounds and compositions. There is, however, another kind of danger, which arises from the fact that anyone can say anything he likes on the Internet. Many websites advertising homeopathic and other kinds of alternative medicine give out false information. It is revelatory to compare the compendium of toxic plants compiled by the European Food Safety Authority (EFSA) with what some of the most popular sites say: whereas this comprehensive inventory makes reference to carefully controlled scientific studies, not a few of the same plants it identifies as toxic are offered for sale by online vendors who tout their benefits for human health.

Among the first plants one comes across in the EFSA compendium is *Aethusa cynapium*, or small hemlock. Some websites advise that this plant (also known as dog poison or fool's parsley) is the source of a homeopathic remedy. The idiocies—very possibly criminal idiocies in some cases—being foisted on an unsuspecting public are quite astonishing. One site claims that the toxic alkaloids this plant contains make it poisonous, but not lethally so; that in homeopathy *Aethusa cynapium* is nevertheless a cure for milk allergy, which causes severe diarrhea in infants; and that it is also effective in reducing intestinal inflammation in adults. Think again.

Next on the list is *Aframomum angustifolium*. Popularly known as Cameroon cardamom, it contains 1,8-cineole, a compound fatal to human beings in doses as small as 0.05 milliliters. After that comes *Aframomum melegueta*, Melegueta pepper or alligator pepper, which contains piperin, an alkaloid found in pepper. Alas, ingesting even 0.35 grams of the seeds of this plant produces blurred or double vision in humans. Then there is a group of fragrant herbs known by the common name *Agastache*. Google this name and one of the first sites to appear, probably devoted to gardening, will tell you something like this: "The agastache plant, also called Mexican hyssop and used in making herbal teas, has a very marked fragrance; crumpled, its leaves give off a quite strong odor similar to mint; its flowers have an odor somewhere between anise, mint, and licorice. In addition to various decoctions that can be made from them, agastache leaves can also be used to flavor raw vegetables, salads, fish sauces, even cakes." The problem is that the agastache plant contains significant amounts of methyl eugenol and estragole, both known carcinogens. Estragole, which is also present in tarragon, basil, anise, and fennel, causes liver damage; in rodents, it is toxic in amounts greater than two grams per kilogram of body weight, but in humans a limit of five grams per kilogram is judged to be acceptable. Neither this information nor the risks of consuming the plant documented by the EFSA compendium are to be found on the website that recommends using it to make teas and to flavor foods. If that website is your only source of information, you may be asking for trouble.

As for essential oils, freely sold over the Internet, where they are typically advertised as "100 percent pure and natural" (hemlock, of course, is natural as well), their estragole content may be 80 percent or higher. Some websites nevertheless assure visitors that these oils are stomachic, diuretic, carminative, antispasmodic, anti-inflammatory, antiviral, and antiallergic, and that they are recommended for the treatment of aerophagia, slow digestion, gastritis and colic, hiccups, spasmophilia, motion sickness, menstrual pain, dysmenorrhea, muscular cramps and contractions, neuritis, sciatica, spasmodic cough, and allergic asthma. What don't they cure?

For better or for worse, few of us are about to give up tarragon or basil just because they contain estragole. Let's move on, then, to *Anacyclus pyrethrum*, Spanish chamomile or Mount Atlas daisy. Hildegarde of Bingen (1098–1179), a German Benedictine abbess whose works include a volume on self-medication, recommended it in these terms: "Spanish chamomile reduces the toxins that contaminate the blood, increases pure blood, and clears the mind. It gives renewed strength to anyone who is weak, indeed failing, and lets nothing leave the body that has not already been digested, but to the contrary assures it a good digestion." The EFSA compendium indicates, however, that the plant's seeds have caused miscarriages in pregnant rats when they are fed for ten days following copulation a dose of 175 milligrams per kilogram of body weight. Fetal malformations were commonly observed as well.

Then there is *Anadenanthera*, a genus of plants containing indolamines derived from tryptamines, among them bufotenine and beta-carbolines. The seeds of its species are hallucinogenic. Specimens may be obtained from websites devoted to ethnobotanical art, for example. One such site cautions visitors that the seeds it sells are for horticultural use only, without, however, explaining why.

Angelica sinensis, also known as dong quai and female ginseng, is recommended by self-styled specialists in naturopathic medicine for people looking to begin a successful weight-loss program. And yet it contains between 0.2 and 0.4 percent essential oils, prominently among them safrole (carcinogenic in rats and mice) and furanocoumarins (or furocoumarins), many of which are toxic not only to insects, but also to mammals.

Yet another example near the head of a very long list is *Annona squamosa*, popularly known as cinnamon apple or sugar apple, an edible green plant with a strange-looking fruit covered by scalelike protuberances. Its pulp is creamy and sweet and, as in other plants of the Annonaceae family, contains a great many black seeds. Eating these seeds and drinking infusions made from the leaves are associated with various types of atypical Parkinsonism. The list goes on. *Artemisia umbelliformis*, white genepi, used in making liquor of various kinds, nonetheless contains alpha- and beta-thujones that have neurotoxic effects. And so on. And on and on.

Some risks are less alarming than others. But they are nonetheless very real. Oxalic acid and oxalates, for example, are toxic substances found in many plants (cocoa, walnuts, hazelnuts, rhubarb, sorel, spinach) that can cause severe local irritation, being easily absorbed by the mucous membranes and the skin, and lead to circulatory problems. When ingested, they can irritate the esophageal and gastrointestinal tracts and cause kidney damage (stones, oliguria, albuminuria, hematuria); in large doses they can be lethal if the precipitation of calcium oxalate into stones blocks the urethra. Again, no one is about to forego a rhubarb tart or salmon with sorel sauce. But nor should anyone suppose that these dishes are completely harmless!

Cooks are therefore well advised to take the warnings in the EFSA compendium seriously and to avoid taking undue liberties. *Borago officinalis*, better known as borage, has an agreeable taste of cucumber, and its flowers have a pleasant oyster flavor. But beware: borage contains toxic unsaturated pyrrolizidine alkaloids. More dangerous still is *Cinnamomum cassia*, China cinnamon, which contains coumarin. The acceptable daily intake of coumarin has been put at 0.1 milligram per kilogram of body weight. A simple calculation shows that this dose is greatly exceeded by a teaspoon of the spice—hence, the importance of distinguishing between China cinnamon and Ceylon cinnamon (*Cinnamomum zeylanicum*), which has a negligibly small amount of coumarin. Coumarin is also present in varying amounts in parsley, celery, parsnips, and bison grass. The use of this last substance as a flavoring agent has led American authorities to prohibit the sale of certain vodkas—much to the dismay of the same connoisseurs who are outraged by the risks associated with pesticide residues in foods.

There's no point working our way through the EFSA compendium much further, I'm afraid. We've gotten only as far as the letter C, and already we have been able to form a fair idea of how irrationally some people will behave even when they have good reason to be cautious. This isn't likely to change any time soon.

SELECTION AND SUPPLY OF COMPOUNDS

The note-by-note cook must have ingredients. Which ones should he use? How can he get hold of them? Each of these questions can be approached in a number of ways. With regard to selection, the outstanding fact, as I say, is that presently there is no list of compounds organized by their molecular characteristics and accompanied by warnings of precautions needing to be taken when they are used for culinary purposes. One could, of course, begin with the more or less comprehensive inventory of traditional ingredients that is available today to food scientists and technologists, who also have at their disposal a great many studies of particular types of compounds (phenolic compounds, fatty compounds, sugar compounds, and so on). The inventory of compounds occurring in traditional ingredients is far from being complete, however, and in the case of fruits, for example, the action of many of the principal compounds is still poorly understood. (Synthetic compounds will never be fully surveyed, by the way, for chemists are adding to their number all the time. One has only to recall the discussion of intense sweeteners in chapter 3 to realize how frequently—and often how fortuitously—such additions are made.) But inventories and studies are almost beside the point. Even more than accurate information, what cooks tempted to experiment with note-by-note cooking need is practical guidance.

Let us therefore distinguish between pure compounds, which in a perfect world note-by-note cooks would use exclusively, and fractions, which may be thought of as analogues of the clusters of musical notes played by synthesizers. Like the basic elements of synthesized music, which do not have to be created one sound wave at a time, vibration by vibration, fractions are composites, combining a variety of compounds that separately stimulate receptors for taste, odor, color, and so on.

PURE COMPOUNDS

Pure compounds are already commercially available from many chemical suppliers. Just as companies that sell laboratory equipment have created special product lines for molecular cooks, companies that produce pure compounds will be able to find niche markets in the culinary field.

The term *pure* should perhaps be put inside quotation marks. Without entering into a lengthy discussion of the practical impossibility of achieving absolute purity, especially at a time when advances in chemical analysis have made it possible to detect the presence of a particular compound in virtually any sample of matter (another reason why the presence of pesticide residues in foods sold in markets should not be condemned out of hand, by the way), let me simply note that chemical suppliers nevertheless are required by law to specify the degree of purity of the compounds they sell. It is common to purchase compounds that are between 95 and 99 percent pure, even more. For dietary purposes, however, these impurities are significant, especially if the compounds in which they occur are used in great quantities (though, of course, any food, whether it is shipped from the farm or created artificially, is essentially water). Compounds sold today for dietary use must meet certain other labeling requirements as well (indicating whether they are food grade, kosher approved, and so on). Looking to the future, one of the chief questions needing to be answered is whether the manufacturers of such products should be allowed to sell them directly to cooks or whether cooks will have to order from a network of suppliers authorized to prepare dilutions under laboratory conditions in regularly inspected plants (like the ones presently operated by wholesale producers of pastry and cured-meat products).

My own view is that if large food companies are permitted to buy additives (by the ton, to be sure) directly from the manufacturer, then restaurants ought to be granted the same privilege. Anyone who claims I have been bought by the additives industry can count on my suing him for slander, by the way, because it is not true. I could easily get rich if I wanted to, but the plain fact of the matter is that I do not run a company that sells compounds for note-by-note cooking, nor do I own any shares in any such company. Nor is it clear, in fact, even considering the prospect of a substantially enlarged market for additives, that companies that currently produce them for the food industry will welcome the upheaval that note-by-note cooking is bound to bring about. For if cooks start to use compounds, what happens to additives? The very notion disappears because in that case every compound becomes an additive—a state of affairs that the present regime of food regulation is utterly inadequate to manage properly.

Legal definitions vary on both sides of the Atlantic. In France, a food additive is officially described as a substance added in small quantities to a food product for a technological purpose. Only those substances that figure in the so-called positive list of additives approved by the European Union and Switzerland (and identified, as we have seen in the preceding chapters, by the letter E followed by a three-number code) can be used for one of several main purposes: lengthening the shelf life of food products by limiting the effects of microbiological contamination (primary and secondary preservatives); lengthening shelf life by limiting the effects of chemical alteration (primary and secondary antioxidants as well as antioxidant enhancers); maintaining or improving physical structure (emulsifiers and texturizers); improving visual appearance (colorants). Today, in France, there are twenty-two approved categories of additives in all: colorants, preservatives, antioxidants, emulsifiers, thickeners, gelling agents, stabilizers, flavor enhancers, acidity correctors, antiagglomerants, modified starches, sweeteners, leavening agents, antifoaming agents, coating agents, emulsifying salts, flour-treatment agents, firming agents, moisturizers, bulking agents, packaging gases, and propellants.

Manufacturers of additives are not the only ones who stand to be affected by the disruptions that note-by-note cooking will cause. Companies that produce and sell OECs will likewise have to face new challenges posed by changing markets and a dramatically altered regulatory environment. Indeed, probably every company today that caters to the food industry will have to make room for small-scale artisanal producers, perhaps even individual entrepreneurs. This is exactly what happened with the advent of molecular cooking. Twenty years ago it was very difficult for chefs to obtain extracts such as agar-agar and carrageenan in the relatively small quantities they needed. At first, they were given free samples. Before long, though, middlemen stepped in, buying in bulk and selling to restaurants on a retail basis. It was a rather straightforward business: one had only to divide tons into kilograms and comply with the relevant legal requirements. Why should it be any different with the various compounds that note-by-note cooks will need?

Even if a product is prohibited by the laws of one country, thanks to the vagaries of electronic commerce it can often be ordered from a foreign source. I do not presume to judge the correctness or desirability of this practice; I merely observe that it is a fact of the world today. It seems equally obvious that governments will soon find themselves very busy monitoring the new markets that are now springing up all over the world as a result. A related question arises as to what will happen if one day individuals take production into their own hands, synthesizing compounds from substances they will have obtained in one way or another by means of a chemical process analogous to caramelization. However small the number of those bold enough to experiment in this fashion may be to begin with, they will need to be properly trained because in order to be certain that the compounds they create are harmless, they have to know exactly what they are doing.

Not long ago researchers in my laboratory were studying what occurs when a particular amino acid, cysteine, is heated in boiling water. What danger could there be in water, you may ask? Isn't the boiling point of water much lower than the temperatures of which even home ovens are capable today? And isn't cysteine found in many of the foods we eat, especially eggs? Correct on all counts—and yet our experiment was accompanied by the release of hydrogen sulfide, a nauseating and toxic gas. As chemists, we were naturally aware of the danger posed by this gas and took the precaution of diffusing it through a solution of sodium hydroxide, trapping it in the form of sodium sulfide, a solid and harmless substance. How many novice note-by-note cooks would know to do this, though?

It must be recognized that molecular transformations do sometimes yield toxic compounds. This is one of the reasons why caramelizing certain artificial sweeteners can be risky, for example. But caramelizing onions and other foods over high heat also needs to be done with care, for the temperatures reached are high enough to produce noxious substances. In recent years, for example, food scientists have conducted many studies on acrylamide, a molecule that is formed during the cooking of products containing flour (breads, pizzas, and so on). Smoking oil can rapidly become sickening as well: the suffocating fumes that rise up from an overheated pan contain acrolein, a very reactive compound that can become attached to molecules in the human organism and disturb bodily functions.

So, yes, note-by-note cooking will call for an abundance of caution. But this should come as no surprise: chemistry is closely related to cooking, both of them having come into existence once mankind learned to harness fire to its own purposes. And we all know you have to be careful when playing with fire!

FRACTIONS

If for one reason or another note-by-note cooks do not wish to deal with pure compounds, there is a simple alternative: to use fractions, obtained for the most part from vegetable and animal products. For several decades now, the food industry has been fractionating wheat and milk, which is to say separating these products into their component parts. In the case of wheat, this method is used to obtain bran, flour, gluten, and starch; with further manipulation, various thickening agents (modified starches, for example) can be derived as well. From milk, fractionation makes it possible to recover fats, lactose, various proteins, and mineral salts. Historically, separating milk fats into subfractions that have various fusing temperatures and then mixing a large amount of easily melted fats with a small amount of fats that melt only at higher temperatures led in turn to industrial production of new dairy products, such as spreadable butter, and new kinds of shortening for pastry making.

More recently, fractionation has been used to extract a variety of products from grape juice and wine, including phenolic compounds, sugars, and odorant fermentation compounds. Modern filtration techniques, together with specially designed membrane filters, make it possible in this case to produce original mixtures that have greater value than the basic raw material itself. After all, why go to the trouble of making thousands of liters of wine when you can produce compounds that will fetch a higher price?

What I am proposing today with note-by-note cooking, in effect, is that fractionation be combined with another chemical technique, cracking, which uses heat to break down complex organic molecules into simpler molecules. Whereas fractionation is a way of separating out the constituent parts of a liquid mixture or solid composite, cracking is a way of modifying the molecular structure of the fractions so obtained. Both of these techniques are well known in petroleum refining, where the hydrocarbons present in the kerosene fraction are heated to yield the smaller hydrocarbon molecules and alkenes used in making gasoline, for example. But they can be applied to any traditional food. Apples are an obvious candidate (in France alone apples are so abundant that the cider industry can't turn all of them into cider), along with carrots, turnips, cauliflower, and, in addition to fruits and vegetables, animals and fish—even insects. Breeding, raising, and harvesting crickets on a massive scale would produce literally tons of protein, an encouraging prospect in view of the challenge of feeding a huge and expanding world population. Few people will want to eat insects, you say? Why not then process them to yield proteinacious fractions that could be used in new and more palatable ways in preparing other dishes?

For the moment, at least, there is no urgent need to go quite that far. But a growing number of companies already manufacture powders, concentrates, and other products from vegetables, fruits, meat, and fish. Until now their wares have been destined mainly for large food-processing firms, but marketing strategy is now beginning to change in response to increasing demand in the culinary sector. Before long, in addition to pure compounds, restaurant chefs and caterers will be able to obtain fractions as well. As in the case of molecular cooking, it may be expected that commercial entrepreneurs, working alone or in small groups, will step forward to act as intermediaries in the supply chain. Indeed, small-scale agricultural producers may be among the first to assume this role. After all, the basic components of a modern filtration system, a pump and a cartridge filter, are inexpensive. Business-savvy farmers will quickly see the point of adding to the value of their products directly at the farm, instead of selling them at ridiculously low prices to supermarkets via food wholesalers. Their example is likely to be followed by note-by-note cooks eager to take advantage of the availability of such equipment, just as molecular cooks acquired rotary evaporators in order to recover odorant fractions from raspberries, coconuts, coffee beans, and so on.

COMPOSITIONS

Let's conclude this market survey—a sort of cook's tour, if you will forgive me yet another play on words—by going back to the comparison with music I mentioned earlier. Synthesizer manufacturers soon came to realize that since composing music one sound wave at a time is too complicated a business for anyone to bother with except physicists and engineers, it was better to give musicians a restricted (though still very large) menu of possibilities in the form of preprogrammed combinations of notes. Hence the introduction and, not long after, the triumph of the modern synthesizer, whose sounds can now be heard everywhere, even in children's toy stores.

It seems logical to suppose that suppliers of products for note-by-note cooking will likewise create specific combinations of compounds—extracts and compositions, as they are known in the trade—to simplify the chef's task. We have already looked at existing industrial preparations of this sort. A new generation of products may carry on from where liquid vanilla extracts, which give the color, the smell, and a bit of the flavor of vanilla, and black and white truffle oils, which do something similar for truffles, left off. All such products are meant to remind us of familiar ingredients. The food-processing industry is aided in its selection of OECs by catalogs in which the number of strawberry flavors, for example, runs into the hundreds! These catalogs are organized not only by flavor type, but also by the type of use to which a particular product is suited (one doesn't use the same OECs in manufacturing aerosol deodorants as in making pastries, for example). The specific purpose that is intended and in particular the type of odor that is desired introduce a further element of complication since particles that are to be dispersed into the air cannot be formulated in the same way as ones that are to be dissolved in a fatty substance. Naturally one hesitates to make very specific predictions about what the future holds, but it would be surprising if the development of note-by-note cooking does not considerably add to the number of suppliers and the number of services presently provided to the culinary sector of the economy.

POLITICAL CONSIDERATIONS

Last, but not least, we come to what might be called the political aspects of note-by-note cooking. Let me make it clear at the outset that I use the word *political* not in the modern sense it has come to acquire from partisan disagreements about who should govern and how, but more generally in the ancient Greek sense of a polis: the public sphere in which members of a polity come together to deliberate on matters of common interest. I touched on this point at the beginning of the book, and it is only fitting that I return to it here, by way of conclusion, in view of its overriding importance. No matter how attractive the promise of note-by-note cooking may seem to true artists, there is a deeper and much broader reason for making the departure from culinary tradition that it represents. It is a way of responding to the concern we all share about the kind of world we will leave to our children. For it is our children and their children who will make up the world of tomorrow, which all of us must hope will be better than the one we live in today.

FROM FARM LAB TO TABLE

Many city dwellers forget that their food comes from agriculture and animal husbandry, and even those who do not are increasingly unaware of what farming today actually involves. Fewer and fewer people now live in the countryside. Agriculture relies to an even greater extent than before on mechanization and chemical fertilizers, which together have made it possible to increase production by an amount that not so very long ago would have been unimaginable. A farmer used to have to work an entire day just to feed four people; today, in the same amount of time, he manages to feed hundreds. And yet not all farmers have as easy a life as city dwellers. Now that there are seven billion of us on earth and counting, we all have an obligation to take an interest in the conditions under which the food so many people depend on is produced. This matter has a number of aspects, bearing not only on production capacity, but also on long-term environmental sustainability and the share of agriculture in a country's gross national product.

The idea that agricultural products might be fractionated at the farm itself seems to me to point to a promising alternative to the present system. It is certainly feasible from the economic point of view (as I say, filtration equipment hardly costs more today than winemaking equipment). If farmers are to be able to sell their fractions, however, it will be necessary, at the other end of the supply chain, that cooks know how to use them. In the coming years, then, if we are to succeed in bringing about a virtuous circle in which all citizens will equitably share the benefits of proper nutrition, both farmers and cooks will need to be trained in the techniques of note-by-note cooking.

With regard to the question of environmental sustainability, so very much on people's minds today, it may be helpful to recall that guano used to be imported to Europe from South America at great cost in order to amend the soil. When sufficient quantities could no longer be found to meet the growing needs of agricultural production, science came to the rescue. The synthesis of nitrates, without which many millions of people would have gone hungry or actually died from starvation, marked a great step forward. No doubt this is why the chemist Fritz Haber (1868–1934) was awarded the Nobel Prize in Chemistry in 1918 for his research on the synthesis of ammonia. There is a terrible irony in this award, for Haber, a German Jew, had also contributed to the development of poison gas for use in chemical warfare during the First World War and later directed research that led to the production of Zyklon B gas, used during the Second World War to kill Jews from every part of Europe. Personally, I find it impossible to agree with the Nobel Committee's decision in this case.

It is important to bear in mind that our present technical capabilities are very limited, at least by comparison with nature itself. Only a short time ago the synthesis of vitamin B12 required the concerted efforts of hundreds of highly skilled chemists. Vegetables, by contrast, spontaneously manufacture this compound with the aid of wind, rain, sun—and billions of years of biological evolution! If we are to make better use of the good Lord's rich storehouse than we have so far managed to do even with modern methods of agriculture and food processing, we will have much yet to learn not only about vegetables, but also about chemical techniques for recovering their compounds. I say nothing about traditional methods of agriculture and land management. What a waste it is to burn forests when, in addition to cellulose, the lignin their trees contain can easily be extracted to make vanillin; when their plants are a source not only of medicines and cosmetics, but also of a great many sapid and odorant compounds that can be used in note-by-note cooking.

As against those who in their ignorance of chemistry believe that it occupies too large a place in our daily lives—wrongly confusing chemistry, which is a science, a branch of knowledge, with its applications—I maintain that what we need is more chemistry, not less; more knowledge, not less, about how atoms can be rearranged and manipulated for the benefit of mankind. It is to chemistry that we must look for instruction if we are to be able to feed ourselves, cure illnesses, and promote health in the years ahead.

REGULATION

People sometimes complain about government agencies responsible for safeguarding public health, ensuring the humane treatment of animals, prosecuting fraud, and so on. But they should pause for a moment and consider how fortunate they are that these agencies exist. Some of us are old enough to remember that toxic cooking oil killed hundreds of people in Spain in 1981. More recently, in 2008, greed led at least one dairy in China to market baby formula contaminated with melamine. In the most highly industrialized countries, where the risk of such behavior has been reduced to minute levels and where even the least threat to public health is immediately and widely publicized, it has been forgotten that only a century ago whole chapters of books on household management were devoted to telling home cooks how to recognize adulterated products. Scandals were common a hundred years ago, when milk was stretched with water, flour was bulked up with plaster, and coffee was flavored with slurry. One guide even recommended inserting a thin metal blade into baby gherkins to check if their green color had been enhanced with copper sulfate (if it had been, the blade would be coated with copper). Imagine how much copper sulfate must have been used back then!

People want to be protected, yet they lash out against the very same public bodies that were created to protect them. As unreasonable as this behavior evidently is, popular dissatisfaction is likely only to increase once note-by-note cooking becomes established, for it will disrupt long-standing business practices and render the laws and regulations pertaining to them obsolete—giving government agencies even more to do than before. In France, debate will certainly take as its point of departure the 1905 law I mentioned earlier, requiring that food products be "wholesome, genuine, and salable."

WHOLESOMENESS

No one will dispute that new food products must be wholesome, even if this idea, taken to a logical extreme, borders on the utopian. After all, even water can be dangerous in large doses. As a practical matter, however, responsible practitioners of note-by-note cooking should not have any great difficulty satisfying this first requirement, at least so long as they rely on the familiar repertoire of proteins, lipids (oils, especially), and carbohydrates.

But even assuming that cooks do their best to take the necessary precautions, there will be accidents if irresponsible manufacturers and suppliers fail to notify them of risks associated with the use of a particular product. Earlier I mentioned the young German molecular cook who suffered horrible injuries from the explosion of a thermos bottle that he had filled with liquid nitrogen and then hermetically sealed, unaware that liquid nitrogen rapidly evaporates at room temperature. It may be confidently predicted as well that diners will be poisoned if chefs use compounds without due regard for toxic risk and the dangers posed by certain kinds of molecular transformation. But, as I say, accidents regularly occur with old-fashioned kitchen knives, whose dangers are known to everyone. Nothing new here, really.

Just so, the manufacture and sale of note-by-note products will have to be regulated in the public interest, albeit in a manner different from what we are used to today since the nature of the products will have changed. Toxicological studies will have to be diligently performed, with special attention to establishing levels of acceptable daily intake.

GENUINENESS AND SALABILITY

Genuineness in the culinary domain, no less than in any other, means that the product being offered for sale is in fact the same one that is purchased. In this connection, we need to distinguish between ingredients, on the one hand, and foods, on the other.

In the case of compounds, certifying authenticity is a straightforward matter, at least to begin with. The compound contained in a flask or other container bearing a label on which a degree of purity is advertised must be pure in the proportion indicated, and, of course, it must be what the label says it is. So far, so good—but things get complicated quickly. We have had occasion to observe more than once in the preceding pages that compounds may assume different forms. Menthol, for example, depending on its form, may or may not smell like mint. Product labeling must therefore be as precise and as detailed as possible.

The conditions in which compounds are stored cannot be neglected. Just as oils must be kept in a cool place, protected from the light in dark bottles (otherwise in nonreactive metal containers to prevent exposure to metal ions), so too care must be taken with certain compounds whose properties are modified by the weather—heat, humidity, and so on. Carotenoids, for example, may undergo isomerization, with the result that their colors change. In the event that the properties of a compound are altered during storage, a question may arise as to which party should be held legally liable, the manufacturer of the product or the purchaser who assumes responsibility for storing it. Fractions, because they often involve traditional (or at least familiar) ingredients, seem to pose few difficulties. Compositions are likely to be more problematic, however, and their status should therefore perhaps be left up to regulators to decide on a case-by-case basis.

The question of what dishes can legally be called presents an opportunity for making useful distinctions. Courts in a number of countries in Europe have upheld the right of restaurants to serve a coq au vin, for example, that has neither a rooster nor any wine, ruling that the dish can be made instead with a hen (or spring chicken) or grape juice or both. The courts are wrong to legitimize false advertising of this sort, I believe, just as they are wrong to allow a commercially produced sauce to be sold under the name "béarnaise sauce" even though it contains neither eggs nor butter. And yet, to be fair, food manufacturers are merely responding to changes in the way people cook today. Prepared mayonnaise is apt more often than not to be a kind of remoulade since it typically contains mustard—again, following the lead of cooks, both in restaurants and at home, who have gradually strayed from canonical recipes out of a taste for experimentation and a desire to be creative. Food manufacturers are not to blame for this, any more than cooks are. Nevertheless the courts were mistaken in my view to rule as they did.

There is no reason why note-by-note cooking should require such adjudication, however. Unless, of course, the names of dishes mention polyphenols that are nowhere to be found in them or unless a gibbs or a wöhler sauce is not what its name says it is. If note-by-note cooks are honest and show proof of due diligence in vetting new techniques and preparations, they should not worry about having to spend time in court that would better be spent in the kitchen.

THE ENERGY QUESTION

It is a matter of common knowledge that we are faced with an energy crisis. Whether fossil-fuel reserves are actually nearing exhaustion or not, a disputed point, prices for both raw materials and finished products have begun to rise. The culinary implications are unsettling. Start from the fact that in 2010, in France alone, cooking in all its various forms consumed the equivalent of roughly 2.4 million tons of oil and gas (not counting the energy used to heat the water used in cooking). Taking into account the fact that traditional cooking methods waste up to 80 percent of the total energy used, the reason for grave concern, if not quite yet for alarm, is obvious. In the winter, of course, the wasted energy is a welcome source of heat, but what about in the summer? Cooking over high heat naturally consumes the greatest amount of energy, but over time the slow reduction of a stock or a sauce amounts to an equally forbidding luxury—and all the more since filtering, properly done, would achieve the same result at much less cost. What is the point of each cook in a restaurant kitchen making his own reductions when consolidating the work of several stations would yield considerable energy savings? What is the point of making aspics from calves' feet when the smallest and the largest kitchen alike could just as well use gelatin? It is no objection to say that very few kitchens make aspics in this way any more. Mountains of veal bones are still boiled every day to make stocks and demi-glaces and the like. Sooner or later the increasing cost of energy is bound to make such practices look like a pointless extravagance.

SOCIAL ACCEPTANCE

Just as molecular cooking aroused widespread resistance as a result of the food neophobia with which the human race is chronically afflicted, for better and for worse, note-by-note cooking will inevitably have more detractors than proponents, at least at first. And then there will be those who cannot make up their minds. One question that is frequently raised, a perfectly legitimate one, is whether this latest form of avant-garde cuisine is meant for everyone or only for a select audience of prosperous epicures who can afford to eat at expensive restaurants. Once again the answer is unequivocal: note-by-note cooking must aim to reach all citizens.

Let me be perfectly candid. The strategy that was devised to popularize molecular cooking and that has been adopted once more in seeking to promote note-by-note cooking is quite unashamedly based on an argument from authority. It is essentially the same strategy that Antoine-Augustin Parmentier used to win acceptance for the potato in France at a time when people refused to eat it—understandably so, since the Academy of Medecine had declared that eating potatoes caused leprosy! Parmentier, a pharmacist by training, served in the French army during the Seven Years' War in Germany. Famine was still common, a lingering consequence of the Little Ice Age that was then drawing to a close in northern Europe. Parmentier observed that German peasants ate potatoes, which, unlike wheat, could be grown in cold weather on poor land. On returning to France, he devoted himself to the task of cultivating them and then, the harder part, getting people to try them. In a brilliant stroke, Parmentier managed to persuade Louis XVI to wear a potato blossom as a boutonniere. If the king liked potatoes, it must be because they were a luxury. Soon they were all the rage among the upper classes. The fashion gradually spread. Parmentier is said to have grown potatoes in a field that had been granted to him in the Plaine des Sablons district of Neuilly-sur-Seine, outside Paris, and arranged to have royal guards posted there. The guards were instructed, however, not to arrest thieves. The attraction of forbidden fruit proved to be enough to make the potato known among the middle and lower classes and, a few decades later, a familiar feature of the culinary landscape in France.

The same idea was quite deliberately placed in the service of molecular cooking through the International Workshop on Molecular and Physical Gastronomy, an annual event first held in Sicily in 1992, which brought together some of the best scientists and chefs in the world. Everything went according to plan. The vogue for the new cooking spread rapidly, especially in the most fashionable restaurants of the world's major cities, where only people with enough money to treat themselves to a meal costing hundreds of dollars per person—the aristocracy of the present day—can think of eating.

Nevertheless the ultimate objective was to bring molecular cooking into the homes of ordinary people. For most families, changing the way they cook makes sense only if it will also save them money. Why buy eggs to make a chocolate mousse, for example, when there is no need for any? The recipe I devised some years ago for Chantilly Chocolate (in which chocolate is heated in water, producing an emulsion that you then whip to obtain a true mousse) proves the point. Why use ten eggs to make meringues for ten people when a single egg is enough to make quarts of the stuff? The educational system in France played a role as well in the promotion of molecular cooking by introducing students at a young age to a more sensible approach to cooking. Workshops on flavor were set up in primary schools, with an emphasis on simple experiments, leading to more advanced study of the science of cooking in middle and high schools.

For note-by-note cooking, the strategy remains the same. Once again young chefs are being invited to practice a modern art; a new style of cooking and eating is being proposed that young people can make their own (and that most older people will detest); and arguments from authority are being deployed with a view one day to making note-by-note cooking seem not merely obvious, but inevitable.

Still, I cannot imagine taking leave of you, my dear friends, with boastful promises of conquering the world—especially when we have been talking all this time about cooking, a joyful activity whose purpose is to bring human beings together in peace, and the radiant future that awaits food lovers everywhere, an age of brilliant new ideas that will provoke astonishment once they have been given sensuous form by artists whose first and foremost concern is to bring happiness to all those who dine at their table.

Everyone, each in his own way—artists, artisans, parents anxious for the future of their children, children eager to learn and to take their own place at the table of human culture—will be able to participate in a grand adventure, a truly new way of cooking that will be more, much more, than a mere fashion or trend. Together, in helping to construct a new way of cooking, we shall create a new way of eating. The future beckons us!

FURTHER
READING

Ames, Bruce N., Margie Profet, and Lois Swirsky Gold. "Dietary Pesticides (99.99% All Natural)." *Proceedings of the National Academy of Sciences (USA)* 87, no. 19 (1990): 7777–7781.

Bastide, N. M., F. H. F. Pierre, and D. E. Corbet. "Heme Iron from Meat and Risk of Colorectoal Cancer: A Meta-analysis and a Review of the Mechanisms Involved." *Cancer Prevention Research* 4, no. 1 (2011): 1–16.

Bittman, Mark. *Food Matters: A Guide to Conscious Eating with More Than 75 Recipes.* New York: Simon & Schuster, 2009.

During, A. "Carotenoid Journey in the Human Body: From Food to Beneficial Effects." In *Proceedings of 13th World Congress of the International Union of Food Science and Technology* (2006). http://dx.doi.org/10.1051/IUFoST:20060758.

European Food Safety Authority. "Compendium of Botanicals Reported to Contain Naturally Occurring Substances of Possible Concern for Human Health When Used in Food and Food Supplements." *EFSA Journal* 10, no. 5 (2012): 2663. http://www.efsa.europa.eu/en/efsajournal/doc/2663.pdf.

Hsieh, C. L., C. C. Peng, Y. M. Cheng, L. Y. Lin, Y. B. Ker, C. H. Chang, K. C. Chen, and R. Y. Peng. "Quercetin and Ferulic Acid Aggravate Renal Carcinoma in Long-Term Diabetic Victims." *Journal of Agriculture and Food Chemistry* 58, no. 16 (2010): 9273–9280.

Krief, S. "Effets prophylactiques et thérapeutiques de plantes ingérées par les chimpanzés: La notion 'd'automédication' chez les chimpanzés." *Primatologie* 6 (2004): 151–172.

Meugnier, E., C. Bossu, M. Oliel, S. Jeanne, A. Michaut, M. Sothier, J. Brozek, S. Rome, M. Laville, and H. Vidal. "Changes in Gene Expression in Skeletal Muscle in Response to Fat Overfeeding in Lean Men." *Obesity* 15, no. 11 (2007): 2583–2594.

Nesslany, F., D. Parent-Massin, and D. Marzin. "Risk Assessment of Consumption of Methylchavicol and Tarragon: The Genotoxic Potential in Vivo and in Vitro." *Mutation Research/Genetic Toxicology and Environmental Mutagenesis* 696, no. 1 (2010): 1–9.

Plaza-Bolanos, P., A. Garrido Frenich, and J. L. Martínez Vidal. "Polycyclic Aromatic Hydrocarbons in Food and Beverages: Analytical Methods and Trends." *Journal of Chromatography A* 1217 (2010): 6303–6326.

Pollan, Michael. "The Food Movement, Rising." *New York Review of Books*, June 10, 2010.

Rietjens, I. M., M. J. Martena, M. G. Boersma, W. Spiegelenberg, and G. M. Alink. "Molecular Mechanisms of Toxicity of Important Food-Borne Phytotoxins." *Molecular Nutrition & Food Research* 49, no. 2 (2005): 131–158.

Rigby, N., and P. James. *Waiting for a Green Light for Health? Europe at the Crossroads for Diet and Disease.* International Obesity Task Force (IOTF) Position Paper. London: IOTF, 2003.

Sharoni, Y. "A Molecular Basis for the Cancer Preventive Activity of Tomato Carotenoids." In *Proceedings of 13th World Congress of the International Union of Food Science and Technology* (2006). http://dx.doi.org/10.1051/IUFoST:20060827.

Tuomisto, H. L., and M. J. Teixeira de Mattos. "Environmental Impacts of Cultured Meat Production." *Environmental Science and Technology* 45, no. 14 (2011): 6117–6123.

APPENDIX

A FEW RECIPES

▼

THE DEVELOPMENT OF MOLECULAR COOKING was accelerated by the publication of recipes that chefs then tested and gradually modified. Here, then, without any further delay, another small step along a road with no end!

Quantities are not always precisely specified in these recipes. They are meant only to give the reader a sense of proportions and some idea of the sorts of things that are possible. All measures are given in metric units, using the conventions of the International System of Units; a conversion guide to U.S. standard units appears at the end of the appendix, together with a brief list of new note-by-note compounds and compositions.

All recipes are mine unless otherwise indicated. I have ordered them alphabetically, avoiding any suggestion as to whether they should be regarded as appetizers or main courses or desserts. Let the imagination of cooks everywhere decide!

RECIPES

☐ APPLE PEARLS, OPALINES, AND LEMON GRANITA
(PIERRE GAGNAIRE)

Preparation of this dish, first served in Hong Kong on April 24, 2009, involves five steps.

WHEY Mix citric acid and cold milk; heat until the mixture curdles, then strain through a damp cloth and let the separated liquid cool.

APPLE PEARLS Use a syringe to inject drops of an aqueous solution containing calcium lactate, artificial green apple flavoring, and the whey into a beaker of water in which sodium alginate has been dissolved using an immersion mixer.

GLUCOSE PÉLIGOT OPALINES Heat water and glucose until they form a brown caramel; scrape it onto a stick-proof surface and let cool; then reduce to a fine powder with a rolling pin. Next, take an indented baking tray with round pockets 6 cm in diameter and sprinkle the powder into the pockets to a thickness of 2 mm and put the tray in a 160°C (320°F) oven. Once baked, remove the péligot disks, let cool, and keep them in a dry place.

LEMON GRANITA Melt some glucose in water, add citric acid and a lemon odorigenic composition (or else a very dilute limonene solution), and freeze. Crush the solid mass with a fork to achieve the consistency of a granita.

ASSEMBLY Coat the bottom of a shallow bowl with the granita, dot with the pearls, and cover with three péligot disks, one on top of the other.

☐ CARAMEL AND PÉLIGOTS

The following elements can be added to various recipes of your own invention:

CARAMEL Heat some sucrose in water until it colors.

GLUCOSE PÉLIGOT Heat some glucose until it colors.

FRUCTOSE PÉLIGOT Heat some fructose until it colors.

LACTOSE PÉLIGOT Heat some lactose until it colors.

✚ CITRIC VAUQUELIN

Put some egg white powder in water (10 percent by mass), then add a teaspoon (5 ml) of citric acid and a soup spoon (10 ml) of glucose. Beat the preparation until stiff and spoon it out into individual "meringues," then heat them in a microwave oven (or not, as you prefer) and serve in dessert glasses.

✚ DEAD LEAF

To 600 ml water add 400 ml ethanol, then 0.0001 g calcium phosphate, 0.001 g sodium phosphate, 0.3 g oenological tannins, 10 g glucose, and a drop of a dilute solution of paraethylphenol (about four parts per million) to give a peat taste.

✚ EFFERVESCENCE

Prepare a solution of glucose (20 g) in water (500 ml), then add ethanol in an amount equal to 20 percent of the total liquid volume, a teaspoon of granulated sugar (sucrose), and a tablespoon of total polyphenols extracted from Syrah grapes.

Just before serving add a mixture of tartaric acid and sodium bicarbonate (two parts tartaric acid for one part of bicarbonate).

✚ FIBRÉ

Leach some wheat flour to extract the starch, in two steps: first, knead the flour with water to make a firm ball; then gently knead the ball in a salad bowl filled with water so that a white powder (the starch) is separated out and you are left with an elastic material (gluten).

Add water and egg white powder to the gluten and feed the mixture through a pasta extruder so that the strips of dough fall into boiling salted water, forming a kind of spaghetti similar to *échaudés*.

After three minutes remove the note-by-note spaghetti, bundle the strands together and arrange them lengthwise in a cake mold, then moisten the bundle with an aqueous solution containing agar-agar (5 percent by mass), monosodium glutamate (as much or as little as you like), chicken flavoring, salt, and glucose. Let set, then cut crosswise into slabs. (Note that other shapes are possible: hollow tubes similar to rigatoni are shown in the color insert)

Serve with a conglomèle made by using gelatin to "glue" together pearls with a liquid center composed of water, glucose, and beta-carotene. (Making pearls is straightforward: see any website devoted to molecular cooking where—as in France today—high school students post the results of their supervised experiments.)

The accompanying sauce is made by heating a cup of water in a pan until it comes to a boil, then adding a leaf of gelatin (previously soaked in cold water) and a cup of neutral oil. Once the liquid has emulsified, add three drops of truffle oil.

◼ NOTE-BY-NOTE BEET SOUFFLÉ WITH AN ORANGE CENTER

This dish was prepared by the members of the Paris chapter of Les Toques blanches internationales as part of the chapter's December 3, 2011, televised charity event.

INGREDIENTS FOR BEET SOUFFLÉ

35 g egg white powder (10 level soup spoons)
900 g water (90 level soup spoons)
40 g beet glaze
300 g powdered sugar
table salt

INGREDIENTS FOR ORANGE SOUFFLÉ

35 g egg white powder (10 level soup spoons)
900 g water (90 level soup spoons)
20 g orange flavoring
300 g powdered sugar
table salt

ASSEMBLY Spray individual ramekins with vegetable oil and sprinkle powdered sugar on top. Coat the bottom of the ramekins with the beet mixture, put a large spoonful of the orange mixture in the center, then cover completely with the beet mixture, filling the ramekins to the top.

COOKING Put the ramekins under the broiler for a few seconds in order to form a light crust, then cook them in a 160°C (320°F) convection oven for four minutes and thirty seconds. Remove from oven and serve at once. Keep in mind that cooking time varies depending on the particular oven, so it may be a good idea to make a test soufflé in order to see if you need a bit more or a bit less than four and a half minutes.

✚ NOTE-BY-NOTE MOZZARELLA

Like the Note-by-Note Beet Soufflé, this dish was prepared by members of the Paris chapter of Les Toques blanches internationales for the December 3, 2011, televised charity event. It consists of three elements, the uppermost of which was inspired by Patrick Terrien's eponymous creation.

TERRIEN For 15 portions: 200 g clarified butter, 200 g water, 30 g powdered milk, 10 g powdered yogurt, 4 g salt, 4 g iota-carrageenan, 1 g agar-agar, and 100 g neutral oil. Strain the clarified butter through a muslin-lined funnel, heat over a medium flame, adding the other ingredients (except the oil), and gently stirring with a whisk. When the first ripples appear, before the liquid begins actually to boil, remove from heat and vigorously whisk in the oil. Then strain once more, pour into a mold, and let cool.

RED JUICE Dissolve some tomato powder in water. Add monosodium glutamate, salt, and ethanol—just as if you were making a Bloody Mary.

ISOMALT TUILE A *tuile* is a traditional French wafer-thin cookie. For the note-by-note version, heat some isomalt, then add a green coloring agent (E160a, for example) and basil flavoring (if possible, an artificial flavoring).

ASSEMBLY Pour the red juice into an elegant glass of your choosing, then place the isomalt *tuile* over the juice and the terrien on top of the *tuile*.

✚ NOTE-BY-NOTE PIE CRUST

Take some corn starch (94 percent amylopectin), add some gluten, and make a dough to be cooked as you would in making an ordinary pie crust. Decorate with fruits of your choice.

✚ NOTE-BY-YOLK

This dish was adapted from a recipe by Frédéric Lessourd, chef-instructor at École Le Cordon Bleu Paris. It resembles a fried egg, but in this case—a livelier variant of the dish shown in the color insert—with an orange-colored yolk and a blue-colored white. Resting on a circular base, it is accompanied by a thin strip of a preparation known as a gauss and napped with a wöhler sauce.

THE "YOLK" In 10 g water, dissolve a pinch of salt, a tablespoon of glucose, and a few drops of white vinegar. Now incorporate egg white proteins (if none are commercially available, make your own by progressively adding salt to egg whites and recovering the precipitate) and heat until the mixture has coagulated. Then add a teaspoon of soy lecithin and whisk in 40 g soybean oil until emulsified. Finish with a dash of beta-carotene.

THE "EGG WHITE" Salt 200 g hot water and dissolve 10 gelatin sheets (leaves) in it along with a drop of an approved blue coloring, then immediately whisk the solution vigorously until you have a considerable quantity of foam. Ladle the foam into bowls lined with transparent plastic film and put them in a cool place so that the foam will gel.

THE BASE Heat corn starch in a dry pan, without any butter or oil, stirring well until it turns a color somewhere between blond and brown—closer to blond, since then it will have a mushroom taste. Put some of this roasted starch in a salad bowl with gluten and water, knead it into a ball, and roll it out with a rolling pin; then cut out disks 7 cm in diameter with a cookie cutter. Cook them on wax paper at 180°C (356°F) for 20 minutes.

THE GAUSS Spread surimi paste in a thin layer (2 mm thickness) on plastic film and then use a brush to apply a layer of fat obtained by cooling olive oil in the refrigerator until it solidifies, depositing a sediment; decant and recover the solid part, then tint it red with an approved food coloring. Let cool, then fold in two again and again until you have a slab 5 cm thick. Freeze until firm enough to slice into very thin parallelepipeds.

THE WÖHLER SAUCE Over heat, melt 100 g glucose and 2 g tartaric acid in 200 g water. Add 2 g polyphenols, bring to a boil, and thicken with corn starch (about a teaspoonful). Off heat, emulsify the recovered fat in the liquid part of the fractionated oil. If you like, add a little diacetyl in solution to some oil, then add a drop of 1-octen-3-ol in solution (20 parts per million) to some oil.

ASSEMBLY Center the orange preparation on the disks of roasted starch. Unmold the gelled blue foam and make a circular opening in the middle so that it can be placed over the orange preparation, flat side facing upward. Set a gauss parallelepiped alongside it at an angle and around the base nap the plate with wöhler sauce.

⊕ POLYPHENOL CARAMEL DISK (PIERRE GAGNAIRE)

Heat 100 g fondant and 70 g glucose until the mixture reaches an internal temperature of 120°C (248°F). Add 3 g polyphenols and continue cooking the sugar to a temperature of 155°C (311°F).

Remove from heat and add 10 g cocoa butter. Transfer to wax paper, shape into a disk, and press down to a thickness of 1 mm.

⊕ PURE BISQUE

The gladius of the squid (or "pen," the feather-shaped internal organ that supports the squid's mantle) is almost pure chitin. Take some gladii and heat in oil. Then add water, cover the pan and simmer for 20 minutes over low heat.

⊕ ROASTED FLOUR COOKIES

Heat some potato starch in a pan, stirring continuously, until it begins to color. Transfer the lightly browned starch to a salad bowl and add sucrose (table sugar), oil, a few drops of water, a soup spoon (10 ml) of egg white powder, and a teaspoon (5 ml) of gluten. Knead into small balls, flatten, and bake for ten minutes at 220°C (425°F).

⊞ TERRIEN

Adapted from a recipe by Patrick Terrien, a chef-instructor at École Le Cordon Bleu Paris.

FOR 15 PORTIONS 200 g whey (the liquid obtained by clarifying butter in this case), 200 g water, 30 g powdered milk, 10 g yogurt powder, 4 g salt, 4 g iota-carrageenan, 1 g agar-agar, and 100 g of a neutral oil such as grapeseed or corn. Combine and add a drop of heptanone solution (0.002 g heptanone dissolved in 100 g oil).

Strain the clarified butter through a muslin-lined funnel. Add the other ingredients (except the oil) while whisking gently over medium heat. When the first ripples appear in the liquid, before it boils, remove from heat and vigorously whisk in the oil. Then strain once more, pour into a mold, and let cool.

Serve with a caramel deglazed with a very dilute sotolon solution or with fibers obtained by pressing and chopping parsley.

METRIC TO STANDARD U.S. MEASURE CONVERSIONS

AVOIRDUPOIS WEIGHT

1 gram (g) = 0.035 ounces
100 grams = 3.53 ounces

LIQUID MEASURE

1 milliliter (ml) = 0.034 fluid ounces

LINEAR MEASURE

1 millimeter (mm) = 0.039 inches
1 centimeter (cm) = 0.39 inches

NEW NOTE-BY-NOTE COMPOUNDS AND COMPOSITIONS

CONGLOMÈLE An artificial fruit or vegetable. Plant tissues are made of various structural fibers (cellulose, pectins, and so on) that are joined to form cell walls. It is possible to retain the organizational principle of natural products while varying the flavor and consistency. If you make alginate pearls, for example, which contain a flavorful liquid, and "glue" them together, you get a conglomèle.

FIBRÉ An artificial meat. Natural meat is made of bundles of muscle fibers held together by connective tissue. Again, retaining the principle while changing the material, we get a fibré.

GAUSS A multilamellar system, analogous to puff pastry, whose structural principle can be applied to a variety of materials. Named in honor of the German mathematician Carl Friedrich Gauss (1777–1855).

GIBBS A system obtained by gelifying an emulsion. Named in honor of the American physical chemist Josiah Willard Gibbs (1839–1903).

LIEBIG Another sort of physically gelified emulsion, made by adding gelatin to a flavorful aqueous solution (tea, coffee, meat stock, fruit juice, wine, beer, and so on), whisking oil into the liquid, and then waiting for the gelatin to gel. Named after the German chemist Justus von Liebig (1803–1873).

MENDELEEV A generalization of the principle underlying infusion, maceration, and decoction in which various materials are combined with various liquids (water, oil, alcohol) at various temperatures. Named in honor of the Russian chemist Dmitri Ivanovich Mendeleev (1834–1907).

PÉLIGOT An extension of the principle underlying caramel, which is obtained by pyrolosis of sucrose. The use of saccharides other than sucrose (glucose, fructose, lactose, and so on) yields materials having properties different than those of caramel. Named in honor of the French chemist Eugène-Melchior Péligot (1811–1890).

THÉNARD A novel system obtained by adding a concentrated ethanol solution to an egg white. The proteins coagulate in the same way as in a poached egg. Named in honor of the French chemist Louis-Jacques Thénard (1777–1857).

VAUQUELIN If you whip an egg white (or a solution of 10 percent ovalbumin in 90 percent water), you get a whipped egg white. When the volume seems to have reached its limit, adding an aqueous solution (orange juice, for example, or coffee or wine or meat stock) and continuing to whip the foam will make it possible to increase the volume considerably (a world record was set in December 2012, when more than forty liters of foam were obtained from a single egg white). If now you cook this foam in a microwave oven, you get a vauquelin, named after the French chemist Louis-Nicholas Vauquelin (1763–1829), an early follower of Antoine-Laurent de Lavoisier (1743–1794), the founder of modern chemistry.

WÖHLER A sauce made with water, polyphenols, tartaric acid, glucose, gelatin, and emulsified oil. Named in honor of the German chemist Friedrich Wöhler (1800–1882), the first to synthesize an organic compound from inorganic ingredients.

INDEX

▼

benzylmercaptan, 21–22
Berchoux, Joseph, 20
Bernardé, Nicolas, 191
beta-carotene, 19, 36, 179
Biffi, Jean-Pierre, 32
bitterness: of beer, 46, 47, 118–19, 145;
    bitterants, 145–46; extracting bitter
    compounds, 144–45
blue coloring agents, 177
Bocuse, Paul, 197
borage (*Borago officinalis*), 209
borax, 126
boric acid, 126
bouillon, 21–22, 80, 91
brain: adaptability of, 202; and contrast
    recognition, 39; golden brain fable,
    49–52; and hardness/softness per-
    ception, 104; and imperfection, 45;
    and music, 189; and taste and odor
    perception, 68, 119, *120*, 172–73; visual
    perception by, 39, 54, 173–74
bread, 8, 64, 132, 167, 213–14
Brillat-Savarin, Jean-Anthelme, 20, 63,
    64
brown coloring agents, 178–79. *See also*
    caramel

calcium lactate, 112, 228
Camaroon cardamom, 207
cancer, 30. *See also* carcinogens
canthaxanthin, 180
capsaicin, 28, 151, 170. *See also* peppers,
    hot
caramel, 135–36, 144, 178–79, 228, 233
Caramel and Péligots, 228
carbohydrates, 132–33, 203. *See also*
    sugars
carboxylic acids, 164

carcinogens: in plants, 207, 208; in
    smoked products, 14, 24, 25, 31; in
    various spices and aromatic plants,
    31, 158–59. *See also* toxicity
Carême, Marie-Antoine, 64, *65*, 191–92,
    196
carmine and carminic acid, 175
carmoisine (red dye), 175, 176, 177
carob gum, 62, 94
carotenes, 179. *See also* beta-carotene;
    carotenoids
carotenoids, 221. *See also* beta-carotene;
    carotenes
carrageenans, 62, 91, 93–94; availability
    of, 64, 212; classification of, *100t*; in
    note-by-note recipes, 231, 234
carrots: color of, 12, 19, 179; cooking with,
    39, 89, *104t*; cultivated vs. wild, 12;
    shape and preparation of, 39; sugars
    in, 12, 92, 133, 134
carvone, 166
cellulose and cellulose derivatives, 97–
    98, 136, 203–4, 219, 235
chamomile, Spanish, 208
Chantilly Chocolate, 224
cheese, 46, *104t*, 127, 164
chemistry: and the analysis of sugars,
    133; benefits of, 134–35, 219; "chem-
    ical" compounds as misnomer, 9–10,
    21; chemical reactions defined, 8, 11;
    chemist's world view, 69; childhood
    experiments in, 127–28; cooking
    compared to, 1–2, 10–11, 214; cook-
    ing informed by, 169; extraction and
    processing methods, 153–55; frac-
    tionation and cracking, 27, 214–15 (*see
    also* fractions and fractionation); his-
    tory of, 121–22, 132–33, 138, 141–43,

guar gum, 62, 94–95
gumarabic (acacia gum), 62, 95, *100t*
gums, 169. *See also specific gums*

Haber, Fritz, 218
hangovers, 138
hardness, 70–72. *See also* solids
head cold, 170
heat: and aspertame, 141–42; compounds
   created by, 9, 11; contrasting with
   cold, *107t*; and density and viscosity,
   73–74, *73t*; energy costs of, 33–34,
   222–23; and foams, 83; food heated by
   chewing, 150; and fruit, 60; and gels,
   87, *100t*; and meat, 9; over- and un-
   der-cooking, 55; and the production
   of POP2, 183; and storage issues, 221;
   and sugars, 64, 70, 134, 135–36; toxins
   released by, 213–14; used in note-by-
   note cooking, 64 (*see also* recipes and
   suggestions); and various compounds,
   100t, 154, 169
hemlock, 206–7
heptanone, 46, 234
heterocycles, 167–68
hexenal, 163
Hildegard of Bingen, 208
Hladik, Claude Marcel, 118–19
honey, 8, 72, 117, 133, 134, 164
Hooke, Robert, 67
Hookean solids, *67*, 67–68
Horowitz, Robert, 142
Hubel, David, 54
hydrocarbons, 161
hydrochloric acid, 121, 127
hydrogen sulfide, 132, 213
hydrolysis, 131–32
hydroxymethylfurfural, 134

hydroxyproline, 17, *130*, 131. *See also*
   amino acids
hyssop, Mexican, 207

imperfection, 45–46
inclusion of objects, 101, *106–7t*. *See also*
   dimension
infusions (tinctures), 154
insects, 134, 215
International Workshop on Molecular
   and Physical Gastronomy, 224
Internet, 206–8
interpenetration of objects, 101, *110–11t*,
   112. *See also* dimension
ionones, 166
ions, 123. *See also* mineral salts
isomalt, 139, 140, 231

jams and jellies: additives, 144; as interpenetration of phases, 78, 84; in note-
   by-note cooking, 88, *100t*; pectin in,
   60–61, 96, 142; tasting, 68
Jensen, Harald, 141
juices: from fruits and vegetables, 58, 80,
   83, 182, 214; from meats, 60, 80 (*see
   also* bouillon)

katemfe fruit, 143
ketones, 163
Kuentz, Aline, 32
Kurti, Nicholas, 5–7, 20–21

lactic acid, 126, 127
lactitol, 136, 139, 140
lactose, 93
La Varenne, François Pierre de, 1, 195
Lavoisier, Antoine Laurent de, 76–77, 236
lead, 14, 124

molecular cooking: critiques of, 63, 179; deconstruction in, *187t*, 191; defined, 20; emblematic techniques, 63–64, 112; equipment used, 63–64, 216; evoking emotion through, 191; gels used, 63–64; history of, 20, 224; learning, 35; obtaining ingredients, 212–13; recipes, 224; thickening and gelling additives used, 99 (*see also* gelling agents; thickening agents); traditional cooking not replaced by, 33

molecular gastronomy, 5, 20

molecules: and compounds, 9, 16–19, *18*; defined, 9; "handedness" of, 128–29; oligomers and polymers, 92 (*see also* polymers); water molecules, 9, 10, 17, *18*, 18–19. *See also specific compounds and substances*

Moles, Abraham, 76

monosaccharides, 92, 93, 94, 132, 136, *137*. *See also* saccharides; *and specific sugars*

monosodium glutamate, 80, 117, 131–32, 144; in note-by-note cooking, 229, 231

mushrooms: 3-octanol's odor reminiscent of, 161; mannitol found in, 140; morels, 185; octenol and, 3–4, 139, 162, 191; pigments in, 179; wholesome vs. poisonous, 15; wild vs. fresh, 3

music: innovations in, 22–23, 33, 186–89, *187–88t*, 204, 216; nature often imitated by, 48; structure and contrast in, 190

myristicin, 31, 156

names of dishes. *See* nomenclature

"natural" foods, 11–12, 14

natural vs. artificial, 11–16, 156–57; artificial as "good" or "bad," 15; artificially reproducing existing foods, 29, 186; coloring agents, 181–82; natural not always best/safest, 205; natural vs. synthetic compounds, 10–11, 15–16, 21, 34, 156–57

neohesperidinedihydrochalcone, 142

neophobia, 24–25

Newtonian liquids, 73–74. *See also* liquids

Nignon, Édouard, 145, 197

nitrate(s), 122, 126, 218. *See also* potassium nitrate

nitrites, 126

nomenclature (naming dishes), 8, 193–200, 221–22, 235–36

non-Hookean solids, 68–69

non-Newtonian liquids, 74–75. *See also* liquids

nonpolyhedral solids, 44–49

Note-by-Note Beet Soufflé with an Orange Center, 230–31

note-by-note cooking: as an art, 32–33 (*see also* cooking as art); attraction of, 32; breadth of possibilities for, 29, 185–86; compared to other arts (*see* music; painting); by consumers, 21; critiques of, 63; disruptive power of, 211–13, 220; economics of, 33–34, 222–23; first note-by-note dish, *13*, 32, 193, 199; health risks of, 30–31 (*see also* safety; toxicity); history of, 5–8, 20–22, 32, 198–99; learning, 4–5, 35–36; and nomenclature (naming dishes), 198–200; not chemistry, 8,

salt (sodium chloride): concentration and taste, 115; forming a paste with, 78; in note-by-note cooking, 147; as preservative, 25; properties of, 36, 143–44; in a sauce, 33; seasoning eggs with, 124–25; structure of, 72, 121, 123; in surimi, 88; uniqueness lost below the molecular level, 17; used in food manufacturing, 26

saltpeter, 123, 126

sapictive receptors, 68, 119; biology of, 28, 119, *120*, 173; importance of stimulating, 28; map incorrect, 115–16. *See also* flavor; taste

sapid compounds, 28, 81, 115, 118–21, *120*

satiety, sensation of, 58–59

sauces: collagen and, 60; in note-by-note cooking, 33–34, 233, 236; properties of, 81; sweet-and-sour sauces, 126; thickening agents, 62–63, 91, 98; thickening meat juices for, 60; wine reduction, 33

Saulnier, Louis, 197

Schlatter, James, 141

Schutz, H. G., 56–58

selection and supply of compounds, 210–17

selenium, 124

"Sens dessus dessous" (dish), 71–72

shape and form, 37–49, 52; dispersion and superposition of objects of various dimensions, *102–11t*; importance of structure, 38–40; nonpolyhedral solids, 44–49; polyhedrons, 40–44, *41*; symmetry, 42, 44–46; topology, 47–48

sheets, *103t*, *105–9t*, *111t*

siphons, 64, 112

small hemlock, 206–7

smell, sense of. *See* odor; odorant compounds; olfactory receptors

smoked foods, 14, 24, 25

sodium alginate: derivation and properties, 91, *100t*; in note-by-note cooking, 83–84, 87, *100t*, 228

sodium bicarbonate (baking soda), 28, 117, 127–28, 178, 180

sodium nitrite, 126

softness, 63–64, 70–72

solids, 66–72; crystalline vs. noncrystalline solids, 69–71, *70*; defined, 66; deformation of, 66–68; in dispersed systems, 77, *77t*; hardness or softness of, 70–72; Hookean solids, *67*, 67–68; non-Hookean solids, 68–69; nonpolyhedral solids, 44–49; polyhedrons, 40–44, *41*; sensation of satiety after eating, 58–59. *See also* shape and form

solubility: water-insoluble compounds, 69; water-soluble compounds, 27, 36, 69, 82

solutions, 145–46

sorbet, 60, *106t*

sorbitol, 88, 136, 139, 140

sotolon, 4

soup, 58–59, 195, 233. *See also* bouillon

sourness, 126–27. *See also* acids

spheres, 44–45, 47. *See also* pearls

spices: aromatic plants, 145, 150, 166; healthful properties of, 151; odorant compounds in, 151, 165, 166, 167, 168; pungent spices, 169, 170 (*see also* pepper, black; peppers, hot); safety and toxicity of, 31, 158, 207; taste-activating compounds in, 4; and trigeminal

hanced with, 126; using without hesitation, 28; in wine, 80, 129

tartrazine, 174

taste, 114–49; acids and, 28, 126–27 (*see also* acids; *specific acids*); bitterness, 46, 47, 118–19, 145–46; duration of, 192–93; evolutionary purpose of, 118–19; of fats, 148–49; four-taste theory incorrect, 114, 115–16, 126, 138; individual variance in perception of, 115–16, 117–18; learning to recognize different tastes, 115, 116–17; matrix effects, 146–48; mineral salts and, 122–23, 125; note-by-note cooking's precision with regard to, 147–48; perception of, 23, 57, 68–69, 114, 117–18, 145, 172–74; sourness, 126–27 (*see also* acids); sweetness, 115–17, 126–27, 144 (*see also* sugars; sweeteners; *and specific sugars*); taste-receptor map incorrect, 115–16; vocabulary for, 114, 117, 118; of water, 122–23; water solubility and, 68, 119. *See also* flavor; sapictive receptors; sapid compounds; trigeminal receptors; *and specific foods and compounds*

terpenes, 151, 165–66

terrien, 231, 234

Terrien, Patrick, 231, 234

terrines, 88, 89

terroir, 27

tetrahedrons, 40, *41*, 42

texture, 56, 59–60. *See also* consistency

texturing agents. *See* gelling agents; thickening agents

thaumatin, 143

thénard, 235

Thénard, Louis-Jacques, 235

thickening agents, 62–63, *90t*, 94–99, 139–40, 212, 214. *See also specific agents*

thymol, 151, 167

tinctures (infusions), 154

Tirel, Guillaume. *See* Taillevent

titanium, 124

titanium dioxide, 78–79, 176, 181

tomatoes, 31, 180, 231

tongue, taste-receptors on, 115–16, 119. *See also* sapictive receptors

Les Tontons flingueurs (*Crooks in Clover*, 1963 film), 88

top notes, 168–69

topology, 47–48, *48*

toxicity: of additives and compounds, 14, 30–31, 124, 125–26; adulterated foods, 219; bitterness and, 118, 145; concentration (dosage) and, 206; of copper sulfate, 93–94; determining ADI, 176–77, 220; of hydrogen sulfide, 132, 213; of intense sweeteners, 141; Internet unreliable regarding, 206–8; of methanol, 139; of mineral salts, 123–24; of odorant compounds, 158–59; of plants and essential oils, 206–9; risk of, when synthesizing compounds, 213–14; of spices, 31, 158, 207, 209; of traditional foods, 14, 24, 31, 99, 209. *See also* carcinogens; safety

trace elements, 124

traditional cooking: creating new dishes, 37, 38–39; defined, 20; dependent on season, land, and regional tradition, 185–86; disappearance feared, 24, 32–33; energy costs of, 33, 222–23; and gelatin, 60; with gels, 63; imparting consistency through, 89

in, 145; used in food manufacturing, 26; water-based gels, 83–84; water content of food, 27, 30, 62; water-soluble compounds, 27, 36, 69, 82
wheat, 26, 27, 214. *See also* flour
whipped cream, 72, 95
whiskey, 6, 21
white alcohol, 138
white genepi, 209
wholesomeness, 14, 89, 157, 220. *See also* nutrition; safety
Wiesel, Torsten, 54
wine: additives, 8, 124, 125–26; adulteration of, 6–8, 14, 124; as aqueous solution, 80; bouquet, 150; color of,

173; compounds used in wine-making, 125, 129, 133; fractionation of, 214; more profitable than grapes, 35; oenological tannins, 181, 229; wine reduction, 33
wintergreen oil, artificial, 152
wöhler (sauce), 233, 236
Wöhler, Friedrich, 121–22, *122*, 152, 236

xanthan gum, 62, 91, 95
Xenakis, Iannis, *188t*, 189
xylitol, 136, 139, 140, 170

yellow coloring agents, 174–77, 179, 182–84
Young, Thomas, 67